**Jayne Kranat**
**Series Editor: Julian Gilbey**

Cambridge International
AS & A Level Mathematics:

# Probability & Statistics 2
Coursebook

CAMBRIDGE
UNIVERSITY PRESS

University Printing House, Cambridge CB2 8BS, United Kingdom

One Liberty Plaza, 20th Floor, New York, NY 10006, USA

477 Williamstown Road, Port Melbourne, VIC 3207, Australia

314–321, 3rd Floor, Plot 3, Splendor Forum, Jasola District Centre, New Delhi – 110025, India

103 Penang Road, #05-06/07, Visioncrest Commercial, Singapore 238467

Cambridge University Press is part of the University of Cambridge.

It furthers the University's mission by disseminating knowledge in the pursuit of education, learning and research at the highest international levels of excellence.

www.cambridge.org
Information on this title:
www.cambridge.org/9781108407342 (Paperback)
www.cambridge.org/9781108633055 (Paperback + Cambridge Online Mathematics, 2 years)
www.cambridge.org/9781108462259 (Cambridge Online Mathematics, 2 years)



First published 2018

20 19 18 17 16 15 14 13 12 11 10 9 8

Printed in Poland by Opolgraf

*A catalogue record for this publication is available from the British Library*

ISBN 978-1-108-40734-2 Paperback
ISBN 978-1-108-63305-5 Paperback + Cambridge Online Mathematics, 2 years
ISBN 978-1-108-46225-9 Cambridge Online Mathematics, 2 years

# Contents

**Series introduction** vi

**How to use this book** viii

**Acknowledgements** x

**1 Hypothesis testing** 1

1.1 Introduction to hypothesis testing 3
1.2 One-tailed and two-tailed hypothesis tests 14
1.3 Type I and Type II errors 17
End-of-chapter review exercise 1 23

**2 The Poisson distribution** 25

2.1 Introduction to the Poisson distribution 27
2.2 Adapting the Poisson distribution for different intervals 33
2.3 The Poisson distribution as an approximation to the binomial distribution 36
2.4 Using the normal distribution as an approximation to the Poisson distribution 40
2.5 Hypothesis testing with the Poisson distribution 43
End-of-chapter review exercise 2 47

**3 Linear combinations of random variables** 50

3.1 Expectation and variance 51
3.2 Sum and difference of independent random variables 56
3.3 Working with normal distributions 61
3.4 Linear combinations of Poisson distributions 64
End-of-chapter review exercise 3 68

**Cross-topic review exercise 1** 70

**4 Continuous random variables** 74

4.1 Introduction to continuous random variables 76
4.2 Finding the median and other percentiles of a continuous random variable 83
4.3 Finding the expectation and variance 88
End-of-chapter review exercise 4 94

**5 Sampling** 99
5.1 Introduction to sampling 102
5.2 The distribution of sample means 106
End-of-chapter review exercise 5 118

**6 Estimation** 120
6.1 Unbiased estimates of population mean and variance 122
6.2 Hypothesis testing of the population mean 127
6.3 Confidence intervals for population mean 131
6.4 Confidence intervals for population proportion 138
End-of-chapter review exercise 6 143

**Cross-topic review exercise 2** 147

**Practice exam-style paper** 153

**The standard normal distribution function** 155

**Answers** 157

**Glossary** 171

**Index** 173

# Series introduction

Cambridge International AS & A Level Mathematics can be a life-changing course. On the one hand, it is a facilitating subject: there are many university courses that either require an A Level or equivalent qualification in mathematics or prefer applicants who have it. On the other hand, it will help you to learn to think more precisely and logically, while also encouraging creativity. Doing mathematics can be like doing art: just as an artist needs to master her tools (use of the paintbrush, for example) and understand theoretical ideas (perspective, colour wheels and so on), so does a mathematician (using tools such as algebra and calculus, which you will learn about in this course). But this is only the technical side: the joy in art comes through creativity, when the artist uses her tools to express ideas in novel ways. Mathematics is very similar: the tools are needed, but the deep joy in the subject comes through solving problems.

You might wonder what a mathematical 'problem' is. This is a very good question, and many people have offered different answers. You might like to write down your own thoughts on this question, and reflect on how they change as you progress through this course. One possible idea is that a mathematical problem is a mathematical question that you do not immediately know how to answer. (If you do know how to answer it immediately, then we might call it an 'exercise' instead.) Such a problem will take time to answer: you may have to try different approaches, using different tools or ideas, on your own or with others, until you finally discover a way into it. This may take minutes, hours, days or weeks to achieve, and your sense of achievement may well grow with the effort it has taken.



In addition to the mathematical tools that you will learn in this course, the problem-solving skills that you will develop will also help you throughout life, whatever you end up doing. It is very common to be faced with problems, be it in science, engineering, mathematics, accountancy, law or beyond, and having the confidence to systematically work your way through them will be very useful.

This series of Cambridge International AS & A Level Mathematics coursebooks, written for the Cambridge Assessment International Education syllabus for examination from 2020, will support you both to learn the mathematics required for these examinations and to develop your mathematical problem-solving skills. The new examinations may well include more unfamiliar questions than in the past, and having these skills will allow you to approach such questions with curiosity and confidence.

In addition to problem solving, there are two other key concepts that Cambridge Assessment International Education have introduced in this syllabus: namely communication and mathematical modelling. These appear in various forms throughout the coursebooks.

Communication in speech, writing and drawing lies at the heart of what it is to be human, and this is no less true in mathematics. While there is a temptation to think of mathematics as only existing in a dry, written form in textbooks, nothing could be further from the truth: mathematical communication comes in many forms, and discussing mathematical ideas with colleagues is a major part of every mathematician's working life. As you study this course, you will work on many problems. Exploring them or struggling with them together with a classmate will help you both to develop your understanding and thinking, as well as improving your (mathematical) communication skills. And being able to convince someone that your reasoning is correct, initially verbally and then in writing, forms the heart of the mathematical skill of 'proof'.

Mathematical modelling is where mathematics meets the 'real world'. There are many situations where people need to make predictions or to understand what is happening in the world, and mathematics frequently provides tools to assist with this. Mathematicians will look at the real world situation and attempt to capture the key aspects of it in the form of equations, thereby building a model of reality. They will use this model to make predictions, and where possible test these against reality. If necessary, they will then attempt to improve the model in order to make better predictions. Examples include weather prediction and climate change modelling, forensic science (to understand what happened at an accident or crime scene), modelling population change in the human, animal and plant kingdoms, modelling aircraft and ship behaviour, modelling financial markets and many others. In this course, we will be developing tools which are vital for modelling many of these situations.

To support you in your learning, these coursebooks have a variety of new features, for example:

- Explore activities: These activities are designed to offer problems for classroom use. They require thought and deliberation: some introduce a new idea, others will extend your thinking, while others can support consolidation. The activities are often best approached by working in small groups and then sharing your ideas with each other and the class, as they are not generally routine in nature. This is one of the ways in which you can develop problem-solving skills and confidence in handling unfamiliar questions.
- Questions labelled as P, M or PS: These are questions with a particular emphasis on 'Proof', 'Modelling' or 'Problem solving'. They are designed to support you in preparing for the new style of examination. They may or may not be harder than other questions in the exercise.
- The language of the explanatory sections makes much more use of the words 'we', 'us' and 'our' than in previous coursebooks. This language invites and encourages you to be an active participant rather than an observer, simply following instructions ('you do this, then you do that'). It is also the way that professional mathematicians usually write about mathematics. The new examinations may well present you with unfamiliar questions, and if you are used to being active in your mathematics, you will stand a better chance of being able to successfully handle such challenges.

At various points in the books, there are also web links to relevant Underground Mathematics resources, which can be found on the free **undergroundmathematics.org** website. Underground Mathematics has the aim of producing engaging, rich materials for all students of Cambridge International AS & A Level Mathematics and similar qualifications. These high-quality resources have the potential to simultaneously develop your mathematical thinking skills and your fluency in techniques, so we do encourage you to make good use of them.

We wish you every success as you embark on this course.

Julian Gilbey
London, 2018

# How to use this book

Throughout this book you will notice particular features that are designed to help your learning. This section provides a brief overview of these features.

**In this chapter you will learn how to:**

- understand the distinction between a sample and a population
- use random numbers and appreciate the necessity for choosing random samples
- explain why a sampling method may be unsatisfactory
- recognise that a sample mean can be regarded as a random variable, and use the facts that

**Learning objectives** indicate the important concepts within each chapter and help you to navigate through the coursebook.

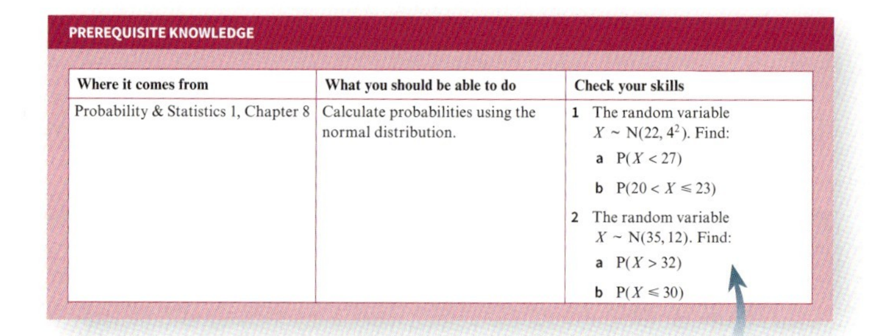

PREREQUISITE KNOWLEDGE

| Where it comes from | What you should be able to do | Check your skills |
|---|---|---|
| Probability & Statistics 1, Chapter 8 | Calculate probabilities using the normal distribution. | 1 The random variable $X \sim N(22, 4^2)$. Find: **a** $P(X < 27)$ **b** $P(20 < X \leqslant 23)$ 2 The random variable $X \sim N(35, 12)$. Find: **a** $P(X > 32)$ **b** $P(X \leqslant 30)$ |

**Prerequisite knowledge** exercises identify prior learning that you need to have covered before starting the chapter. Try the questions to identify any areas that you need to review before continuing with the chapter.

**KEY POINT 1.2**

Data in a stem-and-leaf diagram are ordered in rows of equal widths.

**Key point** boxes contain a summary of the most important methods, facts and formulae.

**WORKED EXAMPLE 5.1**

**a** Show that for samples of size 1 drawn from a fair six-sided die numbered 1, 2, 3, 4, 5 and 6, $E(\bar{X}(1)) = 3\frac{1}{2}$ and $Var(\bar{X}(1)) = \frac{35}{12}$.

**b** Work out $E(\bar{X}(2))$ and $Var(\bar{X}(2))$.

Answer

**a** $E(\bar{X}(1)) = \left(1\times\frac{1}{6}\right)+\left(2\times\frac{1}{6}\right)+\left(3\times\frac{1}{6}\right)+\left(4\times\frac{1}{6}\right)+\left(5\times\frac{1}{6}\right)+\left(6\times\frac{1}{6}\right)$

$= \frac{21}{6} = 3\frac{1}{2}$

You may choose to draw a probability distribution table.

$Var(\bar{X}(1)) = \left(1^2\times\frac{1}{6}\right)+\left(2^2\times\frac{1}{6}\right)+\left(3^2\times\frac{1}{6}\right)+\left(4^2\times\frac{1}{6}\right)$

$+\left(5^2\times\frac{1}{6}\right)+\left(6^2\times\frac{1}{6}\right)-\left(3\frac{1}{2}\right)^2$

$= \frac{91}{6} - \frac{49}{4} = \frac{35}{12}$

**b** $E(\bar{X}(2)) = \frac{1}{2}E(X) + \frac{1}{2}E(X) = E(X) = 3\frac{1}{2}$

$Var(\bar{X}(2)) = \frac{1}{2^2}Var(X) + \frac{1}{2^2}Var(X) = \frac{1}{2}Var(X) = \frac{1}{2}\times\frac{35}{12} = \frac{35}{24}$

You can use expectation algebra, as you have found $E(X)$ and $Var(X)$.

**Worked examples** provide step-by-step approaches to answering questions. The left side shows a fully worked solution, while the right side contains a commentary explaining each step in the working.

distribution is a **Poisson distribution**.

**Key terms** are important terms in the topic that you are learning. They are highlighted in orange bold. The **glossary** contains clear definitions of these key terms.

**EXPLORE 5.4**

Use Microsoft Excel, or the random number function on your calculator, to produce five random single digit numbers. Work out the mean of the five random numbers Repeat this process eight times.

Plot the mean values obtained as a frequency graph, together with the results from each member of your teaching group. What do you notice about the shape of your graph?

What do you think would happen to the shape of the graph if you repeated this process many more times?

What do you think may happen if you had plotted the means of 20 single digit numbers chosen at random instead of 5 random numbers? Try it and see.

**Explore** boxes contain enrichment activities for extension work. These activities promote group work and peer-to-peer discussion, and are intended to deepen your understanding of a concept. (Answers to the Explore questions are provided in the Teacher's Resource.)

**TIP**

A variable is denoted by an upper-case letter and its possible values by the same lower-case letter.

**Tip** boxes contain helpful guidance about calculating or checking your answers.

**DID YOU KNOW?**

Quality control of manufacturing processes is one application of sampling methods. Random samples of the output of a manufacturing process are statistically checked to ensure the product falls within specified limits and consumers of the product get what they pay for. With any product, there can be slight variations in some parameter, such as in the radius of a wheel bolt. Statistical calculations using the distribution of frequent samples, usually chosen automatically, will give information to suggest whether the manufacturing process is working correctly.

**Did you know?** boxes contain interesting facts showing how Mathematics relates to the wider world.

**REWIND**

In Chapter 8 of the Probability & Statistics 1 Coursebook, we learnt that we need to use a continuity correction when approximating a binomial distribution by a normal distribution.

**FAST FORWARD**

We will further explore these results later in this chapter.

**Rewind** and **Fast forward** boxes direct you to related learning. **Rewind** boxes refer to earlier learning, in case you need to revise a topic. **Fast forward** boxes refer to topics that you will cover at a later stage, in case you would like to extend your study.

**Checklist of learning and understanding**

- If $U$ is some statistic derived from a random sample taken from a population, then $U$ is an unbiased estimate for $\Phi$ if $E(U) = \Phi$.
- For sample size $n$ taken from a population, unbiased estimate of:

At the end of each chapter there is a **Checklist of learning and understanding**. The checklist contains a summary of the concepts that were covered in the chapter. You can use this to quickly check that you have covered the main topics.

The **End-of-chapter review** contains exam-style questions covering all topics in the chapter. You can use this to check your understanding of the topics you have covered. The number of marks gives an indication of how long you should be spending on the question. You should spend more time on questions with higher mark allocations; questions with only one or two marks should not need you to spend time doing complicated calculations or writing long explanations.

**END-OF-CHAPTER REVIEW EXERCISE 6**

PS 1 The worldwide proportion of left-handed people is 10%.

a Find a 95% confidence interval for the proportion of left-handed people in a random sample of 200 people from town A. **[3]**

b In town B, there is a greater proportion of left-handed people than there is in town A. From a random sample of 100 people in town B, an $\alpha$% confidence interval for the

**CROSS-TOPIC REVIEW EXERCISE 1**

1 In the manufacture of a certain material, faults occur independently and at random at an average of 0.14 per $1\,\text{m}^2$. To make a particular design of shirt, $2.5\,\text{m}^2$ of this material is required.

a Find the probability that in a randomly selected $2.5\,\text{m}^2$ area of the material there is, at most, one fault. **[3]**

The material is going to be used to make a batch of ten shirts, each requiring $2.5\,\text{m}^2$ of material.

**Cross-topic review exercises** appear after several chapters, and cover topics from across the preceding chapters.

Throughout each chapter there are multiple exercises containing practice questions. The questions are coded:

- PS These questions focus on problem-solving.
- P These questions focus on proof.
- M These questions focus on modelling.
- These questions are taken from past examination papers.

E

**Extension** material goes beyond the syllabus. It is highlighted by a red line to the left of the text.

**WEB LINK**

You can watch a more detailed explanation in the *e* (*Euler's Number*) numberphile clip on YouTube.

**Web link** boxes contain links to useful resources on the internet.

# Acknowledgements

*The authors and publishers acknowledge the following sources of copyright material and are grateful for the permissions granted. While every effort has been made, it has not always been possible to identify the sources of all the material used, or to trace all copyright holders. If any omissions are brought to our notice, we will be happy to include the appropriate acknowledgements on reprinting.*

Past examination questions throughout are reproduced by permission of Cambridge Assessment International Education.

*Thanks to the following for permission to reproduce images:*

Cover image Pinghung Chen/EyeEm/Getty Images

Inside *(in order of appearance)* Paul Bradbury/Getty Images, Science Photo Library, Loop Images/Bill Allsopp/Getty Images, Regie Leong/EyeEm/Getty Images, Universal History Archive/UIG via Getty Images, Bettmann/Getty Images, Douglas Sacha/Getty Images, John Lund/Getty Images, Erik Dreyer/Getty Images, Bettmann/Getty Images, Bettmann/Getty Images, Aeriform/Getty Images, Monty Rakusen/Getty Images, Bettmann/Getty Images, graph by Absolutelypuremilk

# Chapter 1
# Hypothesis testing

**In this chapter you will learn how to:**

- understand the nature of a hypothesis test; the difference between one-tailed and two-tailed tests, and the terms null hypothesis, alternative hypothesis, significance level, critical region (or rejection region), acceptance region and test statistic
- formulate hypotheses and carry out a hypothesis test in the context of a single observation from a population that has a binomial distribution, using:
  - direct evaluation of probabilities
  - a normal approximation to the binomial
- interpret outcomes of hypothesis testing in context
- understand the terms Type I error and Type II error in relation to hypothesis testing
- calculate the probabilities of making Type I and Type II errors in specific situations involving tests based on a normal distribution or direct evaluation of binomial probabilities.

**PREREQUISITE KNOWLEDGE**

| Where it comes from | What you should be able to do | Check your skills |
|---|---|---|
| Probability & Statistics 1, Chapter 7 | Calculate probabilities using the binomial distribution. | **1** $X \sim B(8, 0.1)$. Find:<br>**a** $P(X = 2)$<br>**b** $P(X \leqslant 1)$<br>**c** $P(X > 3)$<br>**2** $X \sim B(15, 0.25)$. Find:<br>**a** $P(X = 7)$<br>**b** $P(X < 2)$<br>**c** $P(2 \leqslant X < 4)$ |
| Probability & Statistics 1, Chapter 8 | Use normal distribution tables to calculate probabilities. | **3** Given that $X \sim N(22, 16)$, find:<br>**a** $P(X < 24)$<br>**b** $P(X > 15)$<br>**c** $P(18 < X < 23)$<br>**4** Given that $X \sim N(30, 6^2)$, find:<br>**a** $P(X < 23)$<br>**b** $P(X > 25)$ |
| Probability & Statistics 1, Chapter 8 | Approximate the binomial distribution to a normal distribution. | **5** Given that $X \sim B(80, 0.4)$, state a suitable approximating distribution and use it to find:<br>**a** $P(X \leqslant 36)$<br>**b** $P(X > 30)$<br>**6** Given that $X \sim B(120, 0.55)$, state a suitable approximating distribution and use it to find:<br>**a** $P(X > 70)$<br>**b** $P(X \leqslant 63)$ |



## Why do we study hypothesis testing?

An opinion poll asks a sample of voters who they will vote for in an election with two candidates, candidate A and candidate B. If 53% of voters say they will vote for candidate A, can you actually be certain candidate A will win the election?

DNA evidence is presented in courtrooms. A judge and/or jury has to make a decision based on this evidence. How can the judge or jury be certain the evidence is true?

A pharmaceutical company testing a new drug treatment observes a positive effect in people using it. How can they test if the effect of the new drug treatment, compared with an older drug treatment, is significant?

These three situations are examples of difficult, real-life problems where statistics can be used to explain and interpret the data. Statistics, and more specifically hypothesis testing, provide a method that scientifically analyses data in order to reach a conclusion. A hypothesis test analyses the data to find out how likely it is that the results could happen by chance. In opinion polls, statisticians look at the sample chosen, whether it is representative, the sample size and how significant the poll results are. Voters looking at opinion poll results may make decisions based on what they read; however, the consequences for individuals are more pertinent in a court case with DNA evidence or when testing new drug treatments. Decisions based on this evidence can be life-changing for individuals.

In this chapter, you will learn how to calculate if the occurrence of events is statistically significant or whether they could have occurred by chance, using the statistical process called hypothesis testing.

**DID YOU KNOW?**

Hypothesis testing is widely used in research in the social sciences, psychology and sociology, in scientific studies and in the humanities. The conclusions reached rely not only on a clear grasp of this statistical procedure, but also on whether the experimental design is sound. To find out further information on experimental design you may like to look at the work of Ronald Fisher, an English statistician and biologist who, in the first half of the twentieth century, used mathematics to study genetics and natural selection.

## 1.1 Introduction to hypothesis testing

Suppose you have a set of dice numbered 1 to 6. Some of the dice are fair and some are biased. The biased dice have bias towards the number six. If you pick one of the dice and roll it 16 times, how many times would you need to roll a six to convince yourself that this die is biased? Write down your prediction of how many sixes imply the die is biased.

Carry out this experiment yourself and work through each step.

Roll your die 16 times and record how many times you get a six. How many did you get? Do you want to reconsider your prediction?

It is possible, although highly unlikely, to roll 16 sixes with 16 rolls of a fair die. If $X$ is the number of sixes obtained with 16 rolls of the die then $X \sim \text{B}\left(16, \frac{1}{6}\right)$ and $\text{P}(X = 16) = \left(\frac{1}{6}\right)^{16}$.

We can calculate theoretical probabilities of rolling different numbers of sixes.

$$P(X=0)=\binom{16}{0}\left(\frac{1}{6}\right)^0\left(\frac{5}{6}\right)^{16}=0.0541 \qquad P(X\leqslant 0)=0.0541=5.41\%$$

$$P(X=1)=\binom{16}{1}\left(\frac{1}{6}\right)^1\left(\frac{5}{6}\right)^{15}=0.1731 \qquad P(X\leqslant 1)=0.2272=22.7\%$$

$$P(X=2)=\binom{16}{2}\left(\frac{1}{6}\right)^2\left(\frac{5}{6}\right)^{14}=0.2596 \qquad P(X\leqslant 2)=0.4868=48.7\%$$

$$P(X=3)=\binom{16}{3}\left(\frac{1}{6}\right)^3\left(\frac{5}{6}\right)^{13}=0.2423 \qquad P(X\leqslant 3)=0.7291=72.9\%$$

$$P(X=4)=\binom{16}{4}\left(\frac{1}{6}\right)^4\left(\frac{5}{6}\right)^{12}=0.1575 \qquad P(X\leqslant 4)=0.8866=88.7\%$$

$$P(X=5)=\binom{16}{5}\left(\frac{1}{6}\right)^5\left(\frac{5}{6}\right)^{11}=0.0756 \qquad P(X\leqslant 5)=0.9622=96.2\%$$

These probabilities tell us that in 16 rolls of a die, almost 89% of the time we would expect to roll at most 4 sixes, and for over 96% of the time we would expect to roll at most 5 sixes. This means that rolling 6 sixes or more will occur by chance less than 4% of the time.

How small must a probability be for you to accept a claim that something occurred by chance? That is, in 16 rolls of the die how many sixes do you need to be convinced the die is biased?



Do you want to reconsider your prediction?

Using the theoretical probability calculations, we can work out $P(X\geqslant 5)$ and $P(X\geqslant 6)$.

$$P(X\geqslant 5)=1-P(X\leqslant 4)=1-0.8866=0.1134=11.3\%$$

$$P(X\geqslant 6)=1-P(X\leqslant 5)=1-0.9622=0.0378=3.78\%$$

The shaded regions on these graphs show the regions for the probabilities. The first graph shows $P(X\geqslant 5)$ and the second shows $P(X\geqslant 6)$.

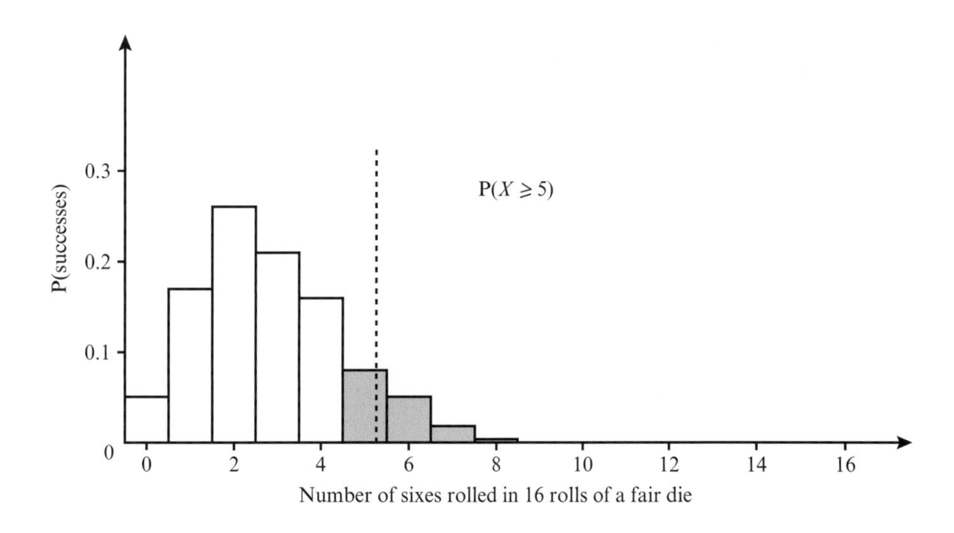

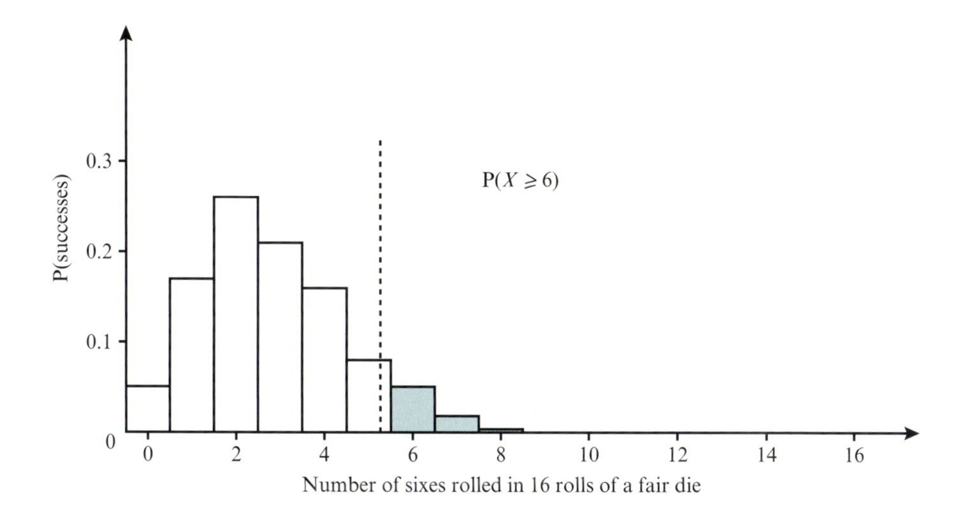


In both graphs, the region to the right of the dotted line represents 5% of the total probability.

The second of the previous two graphs shows that $P(X \geqslant 6)$ is completely to the right of the dotted line, which we would expect as $P(X \geqslant 6)$ is less than 5%. However, the first graph shows the dotted line dividing $P(X \geqslant 5)$ into two parts, which is as expected since $P(X \geqslant 5)$ is greater than 5%. If we choose 5% as the critical percentage at which the number of sixes rolled is significant (that is, the **critical value** at which the number of sixes rolled is unlikely to occur by chance), then rolling 6 sixes is the critical value we use to decide if the die is biased.



The percentage value of 5% is known as the **significance level**. The significance level determines where we draw a dotted line on the graph of the probabilities. This region at one end of the graph is the **critical region**. The other region of the graph is the **acceptance region**.

You can choose the percentage significance level, although in practice 5% is used most often.

### KEY POINT 1.1

The significance level is the probability of rejecting a claim. A claim cannot be proven by scientific testing and analysis, but if the probability of it occurring by chance is very small it is said that there is sufficient evidence to reject the claim.

The range of values at which you reject the claim is the critical region or rejection region.

The value at which you change from accepting to rejecting the claim is the critical value.

### DID YOU KNOW?

This example has links with the philosopher Karl Popper's falsification theory. Popper believed that the claim most likely to be true is the one we should prefer. Consider this example; while there is no way to prove that the Sun will rise on any particular day, we can state a claim that every day the Sun will rise; if the Sun does not rise on some particular day, the claim will be falsified and will have to be replaced by a different claim. This has similarities with the example of trying to decide if the dice is biased. Until we have evidence to suggest otherwise, we accept that the original claim is correct.

Sample data, such as the data you obtain in the dice experiment, may only partly support or refute a supposition or claim. A **hypothesis** is a claim believed or suspected to be true. A **hypothesis test** is the investigation that aims to find out if the claim could happen by chance or if the probability of it happening by chance is statistically significant. It is very unlikely in any hypothesis test to be able to be absolutely certain that a claim is true; what you can say is that the test shows the probability of the claim happening by chance is so small that you are statistically confident in your decision to accept or reject the claim.

**KEY POINT 1.2**

In a hypothesis test, the claim is called the **null hypothesis**, abbreviated to $H_0$.

If you are not going to accept the null hypothesis, then you must have an **alternative hypothesis** to accept; the alternative hypothesis abbreviation is $H_1$.

Both the null and alternative hypotheses are expressed in terms of a parameter, such as a probability or a mean value.

The following example models a hypothesis test.

**WORKED EXAMPLE 1.1**

Experience has shown that drivers on a Formula 1 racetrack simulator crash 40% of the time. Zander decides to run a simulator training programme to reduce the number of crashes. To evaluate the effectiveness of the training programme, Zander allows 20 drivers, one by one, to use the simulator. Zander counts how many of the drivers crash.

Four drivers crash. Test the effectiveness of Zander's simulation training programme at the 5% level of significance.

**Answer**

Let $X$ be 'the number of drivers that crash'.
Then $X \sim B(20, 0.4)$.

$H_0$: $p = 0.4$
$H_1$: $p < 0.4$

Define a **random variable** and its **parameters**. Define the null and alternative hypotheses.

The null hypothesis is the current value of $p$; the alternative hypothesis is training is effective, so $p$ is smaller than 0.4.

The null hypothesis, $H_0$, is the assumption that there is no difference between the usual outcome and what you are testing. The usual outcome is crashing 40% of the time; that is, $p = 0.4$. The alternative hypothesis, $H_1$, is $p<0.4$ since Zander's training programme aims to reduce the probability of crashes.

Four drivers crash; you calculate the test statistic, the region $P(X \leqslant 4)$:

$$P(X \leqslant 4) = \binom{20}{0}0.4^0\,0.6^{20} + \binom{20}{1}0.4^1\,0.6^{19} + \binom{20}{2}0.4^2\,0.6^{18} + \binom{20}{3}0.4^3\,0.6^{17} + \binom{20}{4}0.4^4\,0.6^{16}$$
$$= 0.050952 = 5.10\%$$

The **test statistic** is the calculated probability using sample data in a hypothesis test.

You compare your calculated test statistic with what is the expected outcome from the null hypothesis.

$P(X \leqslant 4) > 5\%$, so 4 is not a critical value.
Therefore, accept $H_0$. There is insufficient evidence at the 5% significance level to suggest that the proportion of drivers crashing their simulators has decreased.

Compare the calculated probability with the percentage significance level. Decide to accept or reject the null hypothesis and comment in context.

In this case, the calculated probability must be less than the percentage significance level to say there is sufficient evidence to reject the null hypothesis.

Your conclusion should be in context of the original question.

Consider this alternative scenario: suppose only three of the 20 drivers crashed after attending Zander's simulator training programme. At the 5% level of significance, what would you conclude?

**Answer**

$$P(X \leqslant 3) = \binom{20}{0}0.4^0\,0.6^{20} + \binom{20}{1}0.4^1\,0.6^{19} + \binom{20}{2}0.4^2\,0.6^{18} + \binom{20}{3}0.4^3\,0.6^{17} = 0.01596$$

$P(X \leqslant 3) = 1.6\% < 5\%$

This is less than the percentage significance level.

At the 5% level of significance there is sufficient evidence to reject $H_0$ and accept $H_1$.

The evidence suggests that the proportion of drivers that crash has decreased and the simulator training programme is effective.

Consider this alternative scenario: suppose four drivers crashed and you were testing the effectiveness of Zander's simulator training programme at the 10% level of significance.

**Answer**

$P(X \leqslant 4) = 5.10\% < 10\%$

As this is less than the percentage significance level, you reject $H_0$ and accept $H_1$. At the 10% significance level you can conclude that there is sufficient evidence to say the simulator training programme is effective.

**EXPLORE 1.1**

Use the simulator training programme example to write out the biased dice experiment described earlier in this chapter, as a hypothesis test. Define the random variable and its parameters for the biased dice experiment.

Define the null and alternative hypotheses. Explain why the default belief is that $p = \frac{1}{6}$.

What are you claiming might be true about $p$?

What do you think would be a sensible size for your critical region?

Calculate the test statistic for your chosen critical region.

What do you conclude?

**KEY POINT 1.3**

When you investigate a claim using a hypothesis test, you are not interested in the probability of an exact number occurring, but the probability of a range of values up to and including that number occurring. In effect that is the probability of a region. The test statistic is the calculated region using sample data in a hypothesis test.

The range of values for which you reject the null hypothesis is the critical region or **rejection region**.

The value at which you change from accepting the null hypothesis to rejecting it is the critical value.

**WORKED EXAMPLE 1.2**

Studies suggest that 10% of the world's population is left-handed. Bailin suspects that being left-handed is less common amongst basketball players and plans to test this by asking a random sample of 50 basketball players if they are left-handed.

**a** Find the rejection region at the 5% significance level and state the critical value.

**b** If the critical value is 2, what would be the least integer percentage significance level for you to conclude that Bailin's suspicion is correct?

**Answer**

Let $X$ be the random variable 'number of basketball players' and $p$ be the probability of being left-handed; then $X \sim B(50, 0.1)$.

Define the distribution and its parameters; this is a binomial distribution as there is a fixed number of trials and only two outcomes.

**a** $H_0$: $p = 0.1$

$H_1$: $p < 0.1$

Significance level: 5%

The rejection region is $X \leqslant k$ such that $P(X \leqslant k) \leqslant 5\%$ and $P(X \leqslant k+1) > 5\%$.

State $H_0$, $H_1$ and the significance level of the test.

| $k$ | 0 | 1 | 2 |
|---|---|---|---|
| $P(X = k)$ | 0.0052 | 0.0286 | 0.0779 |

It can be helpful to show the probabilities in a table.

To calculate the probabilities, use $P(X = r) = \binom{50}{r} 0.1^r 0.9^{50-r}$.

$P(X \leqslant 1) = 0.0338 = 3.38\% < 5\%$

$P(X \leqslant 2) = 0.1117 = 11.17\% > 5\%$

The rejection region is $X \leqslant 1$; the critical value is 1.

Ensure that the question is answered fully.

**b** $P(X \leqslant 2) = 11.17\% < 12\%$, so with a critical value of 2 the significance level needs to be 12% to reject the null hypothesis.

For Bailin to reject the null hypothesis the probability of the region must be less than the significance level.

**KEY POINT 1.4**

To carry out a hypothesis test:

- Define the random variable and its parameters.
- Define the null and alternative hypotheses.
- Determine the critical region.
- Calculate the test statistic.
- Compare the test statistic with the critical region.
- Write your conclusion in context.

**WORKED EXAMPLE 1.3**

A regular pentagonal spinner is spun ten times. It lands on the red five times. Test at the 4% level of significance if the spinner is biased towards red.

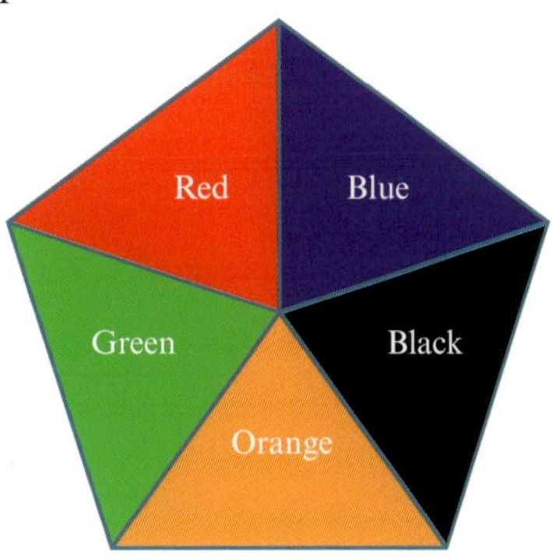


**Answer**

Let random variable $X$ be 'number of spins landing on red'. Then $X \sim B(10, 0.2)$.

Define the random variable and its parameters.

$H_0$: $p = 0.2$.

$H_1$: $p > 0.2$

Define the null and alternative hypotheses.

4% significance level

You are looking for a region with probability less than 0.04.

$P(X \geqslant 5) = 1 - P(X \leqslant 4)$

Calculate the test statistic.

$$= 1 - \left( \binom{10}{0} 0.8^{10} + \binom{10}{1} 0.2\ 0.8^{9} + \binom{10}{2} 0.2^{2}\ 0.8^{8} + \binom{10}{3} 0.2^{3}\ 0.8^{7} + \binom{10}{4} 0.2^{4}\ 0.8^{6} \right)$$

$$= 0.0328 = 3.28\%$$

$3.28\% < 4\%$

Compare with the significance level.

Therefore, reject the null hypothesis. There is evidence to show the spinner is biased towards red.

Interpret the result in the context of the problem.

The graph shows the distribution of the spinner landing on red when it is spun ten times.

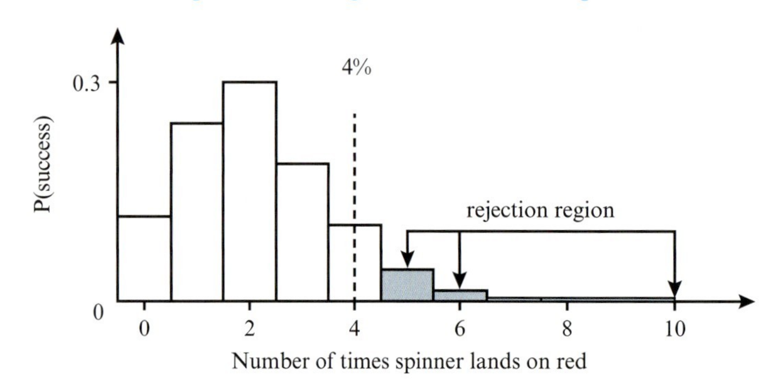


The probability 0.328 or 3.28% is completely in the 4% tail of the graph; that is, the critical, or rejection, region; the null hypothesis is rejected and the alternative hypothesis is accepted.

Note that if the spinner had landed on red four times in ten spins, then $P(X \geqslant 4) = 0.121 = 12.1\% > 4\%$ and the null hypothesis would be accepted; there would not be enough evidence to say the spinner is biased.

**KEY POINT 1.5**

The lower the percentage significance level, the smaller the rejection region and the more confident you can be of the result.



**EXPLORE 1.2**

Why would the significance level be especially important when exploring the data from trials of new medicines?

Discuss the levels of significance you think would be appropriate to use with the following hypotheses.

- Treatment with drug X, which has no known side effects, may not make a surgical option less likely.
- Compound Y, which is expensive, has no effect in preventing colour fading when added to clothes dye.
- Ingredient Z, which is inexpensive, has no effect on the appearance of a food product.

For a large sample, you can approximate a **binomial distribution** to a **normal distribution** and then carry out a hypothesis test.

**REWIND**

In Chapter 8 of the Probability & Statistics 1 Coursebook, we learnt that we need to use a continuity correction when approximating a binomial distribution by a normal distribution.

**WORKED EXAMPLE 1.4**

Arra is elected club president with the support of 52% of the club members. One year later, the club members claim that Arra does not have as much support any more. In a survey of 200 club members, 91 said they would vote for Arra. Using a suitable approximating distribution, test this claim at the 5% significance level.

**Answer**

Let $X$ be 'the number of people who would vote for Arra'. Then $X \sim \text{B}(200, 0.52)$.

The conditions for the binomial are suitable to approximate to a normal distribution.

$\mu = np = 200 \times 0.52 = 104$

$\sigma^2 = npq = 200 \times 0.52 \times (1 - 0.52) = 49.92$

$X \sim \text{N}(104, 49.92)$

When stating the null and alternative hypotheses you can use either the parameter $p$ or $\mu$ but you must be consistent for both $\text{H}_0$ and $\text{H}_1$.

$\text{H}_0$: $p = 0.52$ or $\text{H}_0$: $\mu = 104$

$\text{H}_1$: $p < 0.52$ or $\text{H}_1$: $\mu < 104$

Significance level: 5%

$$\text{P}(X \leqslant 91) = \Phi\left(\frac{91.5 - 104}{\sqrt{49.92}}\right)$$

The test statistic is $\text{P}(X \leqslant 91) \sim \text{P}(X < 91.5)$.

$$= \Phi(-1.769) = 1 - 0.9616$$
$$= 0.0384 = 3.84\%$$

Use normal tables to find the probability.

$3.84\% < 5\%$

Compare with the significance level.

Therefore, reject the null hypothesis. There is some evidence to suggest that Arra does not have as much support as he used to.

Interpret the result.

You may want to use a sketch to check your model and conclusions.

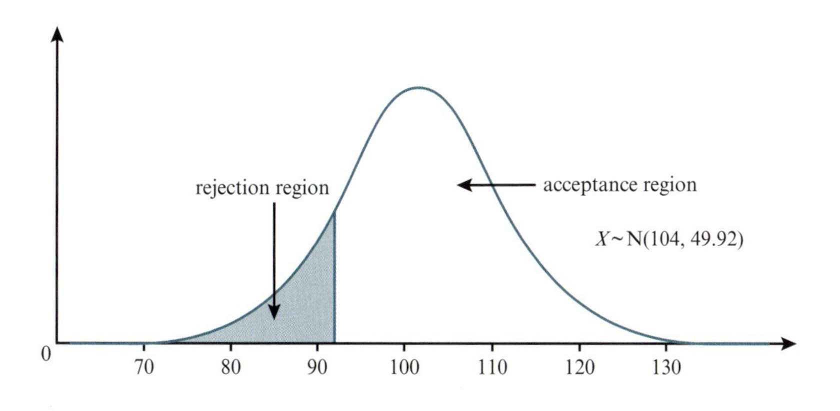




**WORKED EXAMPLE 1.5**

Past records show that 60% of the population react positively to a cream to treat a skin condition. The manufacturers of a new cream to treat this condition claim that more of the population will react positively to the new cream. A random sample of 120 people is tested and 79 people are found to react positively.

**a** Does this provide evidence at the 5% significance level of an increase in the percentage of the population that react positively to this new treatment?

**b** Find the critical value.

**Answer**

a Let $X$ be 'the number of people who react positively to the new cream'. Then $X \sim B(120, 0.6)$.

$np = 120 \times 0.6 = 72$

$npq = 120 \times 0.6 \times (1 - 0.6) = 28.8$

$X \sim N(72, 28.8)$

$H_0$: $p = 0.6$ or $H_0$: $\mu = 72$

$H_1$: $p > 0.6$ or $H_1$: $\mu > 72$

Significance level: 5%

$$P(X \geqslant 79) = 1 - \Phi\left(\frac{78.5 - 72}{\sqrt{28.8}}\right)$$
$$= 1 - \Phi(1.211) = 1 - 0.8871$$
$$= 0.1129 = 11.3\%$$

$11.3\% > 5\%$

Therefore, accept null hypothesis. There is insufficient evidence to support the manufacturer's claim.

The conditions for the binomial are suitable to approximate to a normal distribution.

You can use $p$ or $\mu$ but you must be consistent for both $H_0$ and $H_1$.

Remember to use a continuity correction when approximating a binomial by a normal.

The test statistic is $P(X \geqslant 79) \sim P(X > 78.5)$.

Use normal tables to find the probability.

Compare with the significance level.

Interpret the result.

You may want to use a sketch to check your model and conclusions.

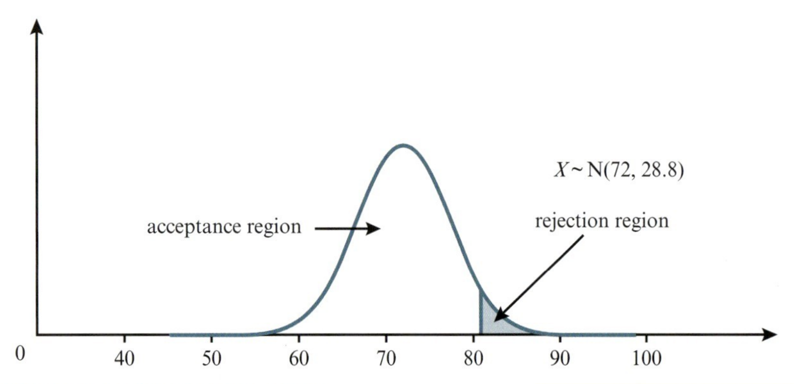


b If $x$ is the critical value, then:

$$1 - \Phi\left(\frac{x - 72}{\sqrt{28.8}}\right) < 0.05$$

$$\Phi\left(\frac{x - 72}{\sqrt{28.8}}\right) > 0.95$$

$$\frac{x - 72}{\sqrt{28.8}} > 1.645$$

$x > 80.8$, so $x = 81$ is the critical value.

The critical value at 5% significance occurs where the probability is less than 5%, or 0.05.

Rearrange to solve for $x$.

Use normal tables in reverse.

**TIP**

When carrying out a hypothesis test, you should always state the conclusion in the context of the question. However, your test is based on a sample of data, so you should not state conclusions in a way that implies that a hypothesis test has proved something; it is better to say 'there is some evidence to show that ...' or 'there is insufficient evidence to show that...'.

**TIP**

As $\Phi(0.75) = 0.674$, then $\Phi^{-1}(0.674) = 0.75$.

## EXERCISE 1A

1 A television channel claims that 25% of its programmes are nature programmes. Yolande thinks the percentage claimed is too high. To test her hypothesis, she chooses 20 programmes at random.

  a If Yolande carries out the hypothesis test at the 10% significance level, define the random variable, state its distribution, including parameters, and define the hypotheses.

  b If there are only two nature programmes, calculate the test statistic. What conclusion does Yolande reach?

  c If the significance level was 5%, what conclusion would Yolande reach?

PS 2 It is claimed that 40% of professional footballers cannot explain the offside rule. Alberto thinks the percentage is lower. To test this, he asks 15 professional footballers to explain the offside rule. Define the random variable and hypotheses. Carry out a hypothesis test at the 5% significance level given that two of the professional footballers asked cannot explain the rule.

PS 3 A farmer finds that 30% of his sheep are deficient in a particular mineral. He changes their feed and tests 80 sheep to find out if the number has decreased.

  a Define the random variable and hypotheses. Using a suitable approximating distribution, carry out a hypothesis test at the 10% significance level given that 19 of the sheep are mineral deficient.

  b Work out the critical value.

PS 4 To test the claim that a coin is biased towards tails, it is flipped nine times. Tails appears seven times.

  a State the critical region and test at the 5% significance level whether the claim is justified.

  b Write out a full hypothesis test of this claim, starting with defining the random variable $Y$ as 'obtaining heads when the coin is flipped'.

  c Dan decides more trials are needed. He flips the coin 180 times and obtains 102 tails. Using Dan's data and a suitable approximating distribution, test the claim at the 5% significance level.

PS 5 To test the claim that a four-sided spinner, coloured red, yellow, green and blue, is biased towards the colour blue, it is spun 120 times. The spinner lands on blue 40 times. If $X$ is the random variable 'the number of times the spinner lands on blue', define the random variable and, using a suitable approximating distribution, test at the 2% level of significance whether the claim is justified.

PS 6 A darts player claims he can hit the bullseye 60% of the time. A fan thinks the player is better than that. Define the random variable and hypotheses. Calculate the test statistic based on the darts player hitting the bullseye 15 times in 17 darts thrown. At what integer percentage significance level is 15 the critical value?



### EXPLORE 1.3

About 8% of the male population is colour-blind. You are doing a survey in which you need to ask one colour-blind male a set of questions. You choose males at random and find that the first 19 males you ask are not colour-blind before you find one who is. Discuss whether you should doubt that 8% of the male population is colour-blind.

## 1.2 One-tailed and two-tailed hypothesis tests

In Section 1.1, the claims made were sure about the direction of the bias; the die was biased towards the number six, Zander was trying to reduce the number of crashes, the spinner was biased towards red. All these examples are **one-tailed** tests where the critical region is at one end, or tail, of the graph.

Consider the example of the spinner. If the claim is that the spinner is not working correctly, rather than the spinner is biased towards a particular colour, then you will apply a **two-tailed** test. This is how you write your working:

Let $X$ be the random variable 'the number of spins landing on red'.

$X \sim B(10, 0.2)$

$H_0: p = 0.2; H_1: p \neq 0.2$

5% level of significance, two-tailed test

$P(X \geqslant 5) = 0.0328$ or 3.28%

3.28% > 2.5%; accept the null hypothesis.

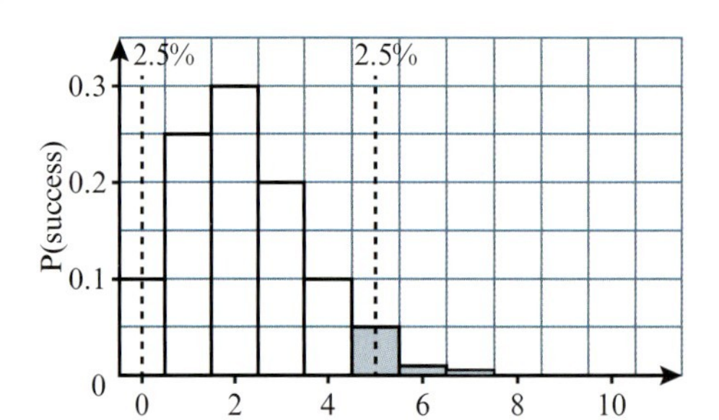


The graph shows the distribution of the spinner landing on red when it is spun ten times.

There is insufficient evidence to support the claim that the spinner is not working correctly.

Note that the alternative hypothesis $H_1$ uses $\neq$ to show that neither an increase nor a decrease is expected; and the significance level percentage has been split to show two critical regions at either end of the graph.



**KEY POINT 1.6**

If a claim is for neither an increase nor a decrease, just a different value of the parameter, then the hypothesis test is two-tailed.

In a two-tailed test use $\neq$ to express the alternative hypothesis and share the significance level percentage between the two tails; that is, between the two critical regions.

Consider Worked example 1.4. If the claim is that Arra's popularity has changed, then you would carry out a two-tailed test. The test statistic gives a probability of 3.84%. At the 5% level of significance, the critical region is shared; half of 5% is 2.5%. Since the probability is found to be 3.84% and 3.84%>2.5%, the null hypothesis is accepted. You would conclude that there is insufficient evidence to suggest that Arra's popularity has changed.

**TIP**

Use the wording in the question to help you decide if your hypothesis test is one-tailed or two-tailed. Look for words that mean increase or decrease for a one-tailed test. For a two-tailed test look for a phrase that implies the parameter is different.

In most two-tailed hypothesis testing the critical region is shared equally between the two tails. However, there may be problems you are modelling when the situation investigated leads you to allocate unequal regions between the two tails.

**WORKED EXAMPLE 1.6**

In a particular coastal area, on average, 35% of the rocks contain fossils. Jamila selects 12 rocks at random from this area and breaks them open. She finds fossils in two of the rocks.

**a** Test at the 10% level of significance if there is evidence to show the average of 35% is incorrect.

**b** In how many of the 12 rocks would Jamilla need to find fossils to reject the null hypothesis?

**c** Would Jamila's conclusion be different if the test was one-tailed?

**Answer**

**a** Let $X$ be the random variable 'the proportion of rocks containing fossils'. Then $X \sim \mathrm{B}(12, 0.35)$.

$\mathrm{H}_0$: $p = 0.35$

$\mathrm{H}_1$: $p \neq 0.35$

Significance level: 10%

Two-tailed test

$$\mathrm{P}(X \leqslant 2) = \binom{12}{0} 0.65^{12} + \binom{12}{1} 0.35^1 \, 0.65^{11} + \binom{12}{2} 0.35^2 \, 0.65^{10}$$

$$= 0.151 = 15.1\%$$

$15.1\% > 5\%$

Therefore, accept the null hypothesis. Jamila must have been unlucky to find so few fossils. She expected to find about four fossils in 12 rocks ($12 \times 0.35 = 4.2$). She found only two fossils, yet this still was too many to provide sufficient evidence to reject the null hypothesis.

Define the distribution and its parameters.

Define $\mathrm{H}_0$, $\mathrm{H}_1$, the significance level and if the test is one-tailed or two-tailed.

Show your working for the test statistic.

Compare with the significance level.

Interpret the result.

You may use a sketch to check your conclusions.

Notice the acceptance region between the two dotted lines.

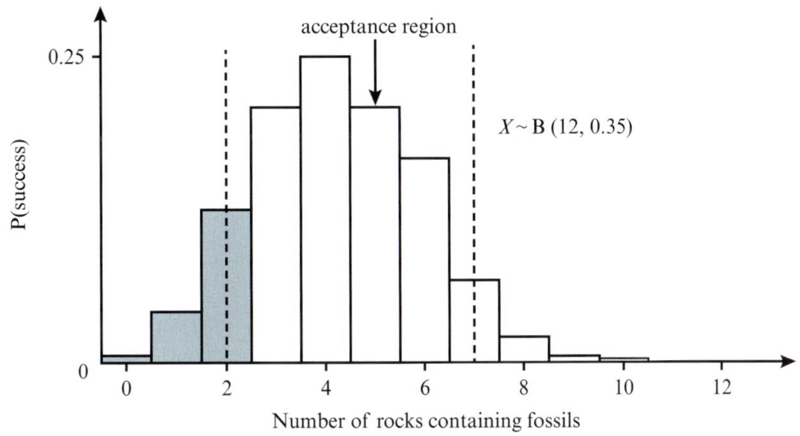

**b** $X \sim \mathrm{B}(12, 0.35)$; $\mathrm{P}(X \leqslant 2) = 0.1513 = 15.13\%$ and $\mathrm{P}(X \leqslant 1) = 0.0424 = 4.24\%$.

Since $\mathrm{P}(X \leqslant 2)$ is greater than 5% and $\mathrm{P}(X \leqslant 1)$ is less than 5%, the critical value is 1.

> $X = 1$ is the critical value for the test.

$\mathrm{P}(X \geqslant 7) = 0.0846 = 8.46\%$ and $\mathrm{P}(X \geqslant 8) = 0.0255 = 2.55\%$

Since $\mathrm{P}(X \geqslant 7)$ is greater than 5% and $\mathrm{P}(X \geqslant 8)$ is less than 5%, 8 is also a critical value.

> In a two-tailed test there are two critical values.

**c** Since $\mathrm{P}(X \leqslant 2) = 15.1\% > 10\%$, the answer is still to accept the null hypothesis: there is insufficient evidence to show that the average number of fossils is incorrect.

> For a one-tailed test the question becomes 'Is there any evidence at the 10% significance level to suspect that the percentage of rocks containing fossils is lower than 35%?'.

## EXERCISE 1B

1 A television channel claims that 25% of its programmes are nature programmes. Xavier thinks this claim is incorrect. To test his hypothesis, he chooses 20 programmes at random and carries out a hypothesis test at 10% significance level

**a** Define the random variable, stating its parameters, and write down the two hypotheses.

**b** If there are only two nature programmes, calculate the test statistic. What conclusion does Xavier reach?

2 It is claimed that 40% of professional footballers cannot explain the offside rule. Zizou does not believe the percentage is correct. To test the claim, he asks 15 professional footballers to explain the offside rule. Define the random variable and hypotheses. Carry out a hypothesis test at the 5% significance level given that two of the professional footballers cannot explain the rule.

3 It is claimed that 80% of Americans believe in horoscopes. Madison doubts the claim is correct.

**a** Madison asks 16 Americans whether they believe in horoscopes. Define the distribution. Using a 10% significance level, write down two hypotheses and test the claim, given that 15 of the people asked believe in horoscopes. What is the critical value?

**b** Madison conducts a larger survey and asks 160 Americans if they believe in horoscopes. Define the distribution. Using a suitable approximating distribution and a 5% significance level, write down two hypotheses and test the claim given that 116 of the people asked believe in horoscopes.

4 It is claimed that 12% of people can correctly spell the word 'onomatopoeia'. Magnus does not believe the claim is correct. He asks 180 people to spell onomatopoeia.

**a** Define the distribution using a suitable approximating distribution and write down two hypotheses. Magnus finds 13 of the 180 people can spell onomatopoeia correctly and carries out a hypothesis test. If 13 is the critical value, work out the percentage significance level and state Magnus's conclusion.

**b** What would Magnus's conclusion be if his belief is that the percentage of people who can spell onomatopoeia is lower than 12%?

5 A manufacturer sells bags of 20 marbles in mixed colours. It claims that 30% of the marbles are red. Ginny thinks this is incorrect and tests the claim by opening a bag of marbles and counting how many are red.

a Carry out a hypothesis test at the 10% level of significance given that Ginny finds three red marbles. Find the critical value for this test.

b Suppose Ginny thinks the percentage should be lower than 30%. State the hypotheses. Will Ginny's conclusion change?

6 It is claimed that the proportion of people worldwide with mixed-handedness (people who swap between hands to perform different tasks) is 1%. Amie thinks this proportion is incorrect. She interviews 600 people, and finds that 11 of them are mixed-handed. Test the claim at the 4% significance level.

**EXPLORE 1.4**

A hypothesis test involves a null hypothesis, an alternative hypothesis, a level of significance, calculation of the test statistic and a conclusion. This may appear to be a straightforward algorithm for reaching a decision about whether results are statistically significant; however, examples used in textbook explanations traditionally give the data alongside the hypotheses and a problem to be addressed.

Discuss the following questions and explore whether the results of using this algorithm to answer hypothesis test questions are statistically meaningful.

- Were the collected data chosen at random? Why is this important?
- Were the data collected before the test was set up, or after the test was set up? Why is this important?
- Why is it important to decide upon the significance level *before* collecting the data?

Write an algorithm for a statistician to follow when setting up a hypothesis test.

## 1.3 Type I and Type II errors

We determine a population parameter from a complete set of data from that population, whereas a sample statistic is calculated from a subset of data from the complete population. When making inferences about a sample of data, such as conclusions from hypothesis testing, we use the sample statistic as if it is a known population parameter. The null and alternative hypotheses are set up with reference to a population parameter and conclusions made from calculations based on a sample of observations. Consequently, there are two ways to reach an incorrect decision based on the results of calculating probabilities in hypothesis testing, a **Type I error** and a **Type II error**. The table shows how these errors arise.

| | **Accept $H_0$** | **Reject $H_0$** |
|---|---|---|
| **The null hypothesis, $H_0$ is true** | Correct decision | Type I error |
| **The null hypothesis, $H_0$ is false** | Type II error | Correct decision |

For example, in England and many other countries the law states that a person is innocent until 'proven' guilty. A Type I error happens when an innocent person is found guilty; a Type II error happens when a guilty person is found innocent. In both these situations, the evidence was either incorrect or not interpreted correctly.

**EXPLORE 1.5**

In the example presented earlier in Section 1.1 – that is, the experiment to determine whether a die is biased – a Type I error occurs when we conclude the die is biased given that it is fair; and a Type II error occurs when we conclude the die is fair given that it is biased. So for a Type I error we would calculate Probability (biased | fair) and for a Type II error we would calculate Probability (fair | biased). Explain the difficulty in calculating P(fair | biased).

In statistics, an error may be made if:

- there is bias in the sample data
- the probability model is not the correct model to use
- the chosen significance level is not appropriate for the situation.

These two diagrams represent the same probability model. The shaded area in each graph represents the acceptance region for a two-tailed test with different significance levels.

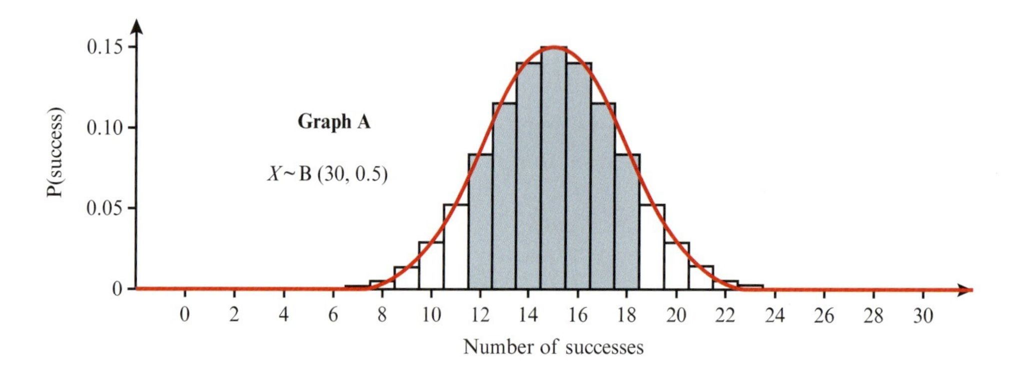




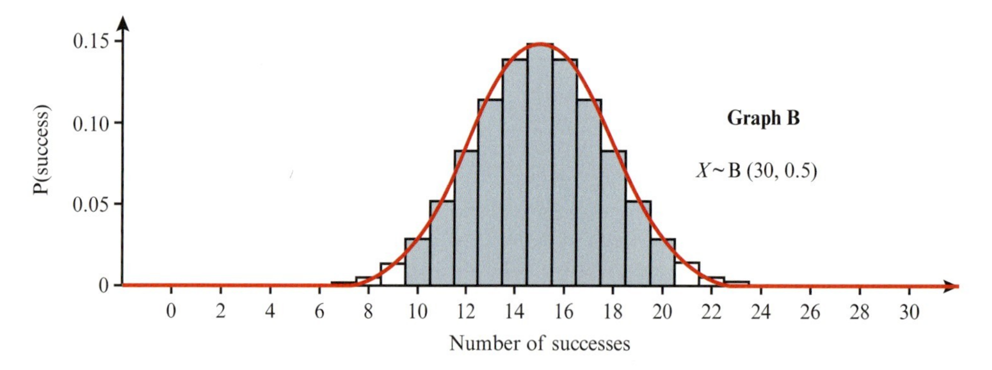


The percentage significance level shown in graph A is higher than the percentage significance level shown in graph B. There is a greater chance of making a Type II error with the situation in graph B because the rejection region – that is, the critical regions in each tail – is quite small, leading us to accept $H_0$ when it is, in fact, false. However, simply changing the significance level to a higher percentage, as in graph A, increases the chance of a Type I error.

**EXPLORE 1.6**

Discuss why the significance level of a test is equal to the probability of making a Type I error.

**WORKED EXAMPLE 1.7**

A football referee uses the same coin for the coin toss at the start of every game he referees. Karam suspects the coin is biased towards heads. He decides to record the result of each coin toss for the next 16 matches, and carry out a hypothesis test at a 5% level of significance.

**a** Find the probability of a Type I error and state what is meant by a Type I error in this context.

**b** Explain why a Type II error is possible if there are 11 tails.

**Answer**

Let $X$ be the random variable 'getting a tail'. Then $X \sim B(16, 0.5)$. $H_0: p = 0.5 \quad H_1: p > 0.5$

Define the random variable.

One tailed test at 5% level of significance.

Next, calculate probabilities using $\binom{n}{t} p^t (1-p)^{n-t}$. You are looking at cumulative probabilities in the upper tail.

The table shows the probabilities of possible outcomes.

| **Number of tails, $t$** | 16 | 15 | 14 | 13 | 12 | 11 |
|---|---|---|---|---|---|---|
| **$P(X \geqslant t)$** | 0.000015 | 0.0003 | 0.0021 | 0.0106 | 0.0384 | 0.1051 |

**a** $P(X \geqslant 12) = 3.84\%$ and $P(X \geqslant 11) = 10.5\%$.

A Type I error occurs if there are 12 or more tails, the probability of which is 0.0384.

In this context, a Type I error means the evidence leads you to suspect the coin is biased when in fact it is not.

A Type I error assumes $H_0$ is true: you would reject $H_0$ where the probability is less than 5%.

The explanation must be given in the context of the question, without mentioning $H_0$ or $H_1$.

**b** $P(X \geqslant 11) = 10.5\%$

$H_0$ would be accepted if there were 11 tails, and if $H_0$ is false this would result in a Type II error.

Ensure you explicitly show results of calculations you are using in your conclusions.

**WORKED EXAMPLE 1.8**

A blood test is used to determine the level, $X$, of a certain protein to determine if a patient has a specific condition. $X$ is abnormally high for people with this condition. Records kept over a long period show that for patients with this condition $X \sim N(22, 7^2)$, whereas for people without this condition $X \sim N(10, 3^2)$.

**a** In a screening test, a doctor decides that anyone with $X > 17$ has the condition. If the null hypothesis is *a person tested does not have the condition*, state the alternative hypothesis and the critical region, and find the probability of the occurrence of:

**i** a Type I error **ii** a Type II error.

**b** Find the range of values of $X$ such that the probability that the test does not correctly identify a person who has the condition is less than 0.03.

**c** Find the value of $X$ where the probability of a Type I error is equal to the probability of a Type II error.

**Answer**

If $H_0$: a person tested does not have the condition, then $H_1$: a person tested does have the condition.

A two-tailed test is appropriate as people either do or do not have the condition.

Reject $H_0$ if $X > 17$.

**a i** If $H_0$ is true, then $X \sim N(10, 3^2)$.

Type I error: $H_0$ rejected when $H_0$ is true.

$$P(\text{Type I error}) = 1 - \Phi\left(\frac{17-10}{3}\right)$$

$$= 1 - \Phi(2.333) = 1 - 0.9902 = 0.0098$$

**ii** If $H_1$ is true, then $X \sim N(22, 7^2)$.

Type II error: $H_0$ accepted when $H_1$ is true.

$$P(\text{Type II error}) = \Phi\left(\frac{17-22}{7}\right)$$

$$= \Phi(-0.714) = 1 - 0.7623 = 0.238$$

**b** $P(X < a) < 0.03$

This is a Type II situation. Use $X \sim N(22, 7^2)$ and normal tables to find $z$ value for $1 - 0.03$.

$$\Phi\left(\frac{a-22}{7}\right) < 0.03$$

$$\frac{a-22}{7} < -1.881$$

$$a < 8.83$$

**c** Let the value be $x$. Then:

For Type I error use $X \sim N(10, 3^2)$.
For Type II error use $X \sim N(22, 7^2)$.
$1 - \Phi\left(\frac{x-10}{3}\right) = \Phi\left(\frac{10-x}{3}\right)$

$$1 - \Phi\left(\frac{x-10}{3}\right) = \Phi\left(\frac{x-22}{7}\right)$$

$$\Phi\left(\frac{10-x}{3}\right) = \Phi\left(\frac{x-22}{7}\right)$$

$$x = 13.6$$



**EXERCISE 1C**

1 An archer claims that her skill at firing arrows is such that she can hit the bullseye 40% of the time. A hypothesis test at the 5% significance level is carried out to test if the archer is as good as she claims to be. The archer fires 12 arrows at a target.

**a** Define the random variable and its parameters.

**b** State the null and alternative hypotheses.

c Find the rejection region.

d Show that a Type I error occurs if the archer hits the bullseye just once.

2 A supplier of duck hatching eggs claims that up to 80% of its duck eggs will produce ducklings. Gareth orders 16 duck hatching eggs and carries out a hypothesis test at the 10% significance level to test this claim.

a Define the random variable and its parameters.

b State the null and alternative hypotheses.

c Find the rejection region.

d Find the probability of a Type I error.

3 A chocolate manufacturer claims that only one-fifth of teenagers can tell the difference between its chocolate and the leading brand. Priya suspects that the claim is too high. For a random sample of 300 teenagers, at a 5% significance level, determine the number that would need to tell the difference for a Type I error to occur. In your working use a suitable approximating distribution and state the null and alternative hypotheses. Remember to use a continuity correction.

4 A gardener knows from experience that the height, $X$, of his sunflowers is normally distributed with mean 1.8 m and standard deviation 0.32 m. A friend claims that she can make sunflowers taller by singing to them daily. They decide to test the claim. One plant, selected at random, is sung to each day while it grows.

a State the null and alternative hypotheses.

b The gardener will be convinced that singing increases sunflower height if $X > 2.25$ m. Calculate the probability of a Type I error.

c The friend says that the mean height of sunflowers sung to is 2.4 m. Assuming the standard deviation is the same, calculate the probability of a Type II error.

5 Janina has an eight-sided die, with sides numbered 1 to 8, which may or may not be biased towards the number 1. The die is rolled 20 times and Janina decides that if the number 1 appears four times, the die must be biased.

a Explain what is meant by a Type I error in this context. Find the probability of a Type I error.

b State what is meant by a Type II error in this context. What additional information is needed to calculate the probability of a Type II error?

## EXPLORE 1.7

A medical researcher conducts a hypothesis test to compare the effectiveness of two medicines used to treat a certain condition.

The null hypothesis $H_0$: the two medicines are equally effective.

The alternative hypothesis $H_1$: the two medicines are not equally effective.

Discuss:

- What conclusions will the researcher reach if a Type I error is made? What are the implications of a Type I error for people who require treatment for this condition?
- What conclusions will the researcher reach if a Type II error is made? What are the implications of a Type II error for people who require treatment for this condition?
- Are the consequences of making one type of error more life-threatening than making the other type of error? How can you address this when determining the significance level?

### FAST FORWARD

Hypothesis testing is an essential process used in data-based decision-making. You will carry out hypothesis testing when modelling problems using other **probability distributions** in Chapter 2 and for other situations using the normal distribution in Chapter 6.

## Checklist of learning and understanding

- To carry out a hypothesis test:
  - Define the random variable and its parameters.
  - Decide if the situation is one-tailed or two-tailed.
  - Define the null and alternative hypotheses.
  - Decide on the significance level.
  - Calculate the test statistic.
  - Compare the test statistic with the critical value.
  - Interpret the result and write a conclusion in context.
- A Type I error occurs when a true null hypothesis is rejected.
- A Type II error occurs when a false null hypothesis is accepted.

## END-OF-CHAPTER REVIEW EXERCISE 1

PS **1** A potter making clay bowls produces, on average, four bowls with defects in every 25. The potter decides to use a new type of clay.

**a** For a randomly chosen sample of 25 bowls made using the new clay, one bowl has a defect. Test at the 5% level of significance if there is evidence that the new clay is more reliable than the old clay. **[5]**

**b** For quality control, a larger batch of bowls is checked for defects, and in a random sample of 150 bowls, defects are found in 18 bowls. Using this larger batch, test at the 5% level of significance if there is evidence that the new clay is more reliable than the old clay. **[5]**

M **2** Rainfall records over a long period show that the probability that a particular area of low-lying land will be flooded on any day is 0.2. A meteorologist suspects that, due to climate change, the land will be flooded more often. Over a randomly chosen set of 1000 days from recent years, the area of low-lying land was flooded on 218 days. Investigate at the 5% level of significance whether or not the meteorologist's suspicion is justified. **[5]**

M **3** A manufacturer claims that the life of light bulbs it produces is normally distributed with mean 800 hours and standard deviation 50 hours. You test two light bulbs to destruction and find that one lasts for 720 hours and the other lasts for 920 hours.

**a** Determine, at the 5% significance level, if either 'test to destruction' provides evidence for the manufacturer's claim and comment on your conclusion if you were to use a light bulb from this manufacturer. **[5]**

**b** John purchases a light bulb and is only concerned if the light bulb does not last for as long as is claimed. Determine, at the 2.5% significance level, how long, to the nearest hour, the light bulb would last for a Type I error to occur. **[5]**

PS **4** A questionnaire has 40 questions. Responses to this questionnaire are recorded as a single digit, either 0 or 1. For each digit, the probability that it is recorded incorrectly is 0.08. For a one-tailed test, at the 10% significance level, find the critical region and state the critical value. **[5]**

PS **5** It is claimed that the proportion of people worldwide with red hair is 1.2%. Margaret thinks the proportion of people in Scotland with red hair is higher. At her school there are 850 students; 15 of the students have red hair.

**a** Test Margaret's claim at the 2% significance level. **[5]**

**b** Find the critical value. **[3]**

 PS **6** It is claimed that 30% of packets of Froogum contain a free gift. Andre thinks that the actual proportion is less than 30% and he decides to carry out a hypothesis test at the 5% significance level. He buys 20 packets of Froogum and notes the number of free gifts he obtains.

**i** State null and alternative hypotheses for the test. **[1]**

**ii** Use a binomial distribution to find the probability of a Type I error. **[5]**

Andre finds that three of the 20 packets contain free gifts.

**iii** Carry out the test. **[2]**

*Cambridge International AS & A Level Mathematics 9709 Paper 73 Q5 November 2016*

**7** Marie claims that she can predict the winning horse at the local races. There are eight horses in each race. Nadine thinks that Marie is just guessing, so she proposes a test. She asks Marie to predict the winners of the next ten races and, if she is correct in three or more, Nadine will accept Marie's claim.

**i** State suitable null and alternative hypotheses. **[1]**

**ii** Calculate the probability of a Type I error. **[3]**

**iii** State the significance level of the test. **[1]**

*Cambridge International AS & A Level Mathematics 9709 Paper 73 Q2 June 2015*

8 A machine is designed to generate random digits between 1 and 5 inclusive. Each digit is supposed to appear with the same probability as the others, but Max claims that the digit 5 is appearing less often than it should. To test this claim the manufacturer uses the machine to generate 25 digits and finds that exactly 1 of these digits is a 5.

i Carry out a test of Max's claim at the 2.5% significance level. **[5]**

ii Max carried out a similar hypothesis test by generating 1000 digits between 1 and 5 inclusive. The digit 5 appeared 180 times. Without carrying out the test, state the distribution that Max should use, including the values of any parameters. **[2]**

iii State what is meant by a Type II error in this context. **[1]**

*Cambridge International AS & A Level Mathematics 9709 Paper 73 Q6 June 2014*

9 A cereal manufacturer claims that 25% of cereal packets contain a free gift. Lola suspects that the true proportion is less than 25%. To test the manufacturer's claim at the 5% significance level, she checks a random sample of 20 packets.

i Find the critical region for the test. **[5]**

ii Hence, find the probability of a Type I error. **[1]**

Lola finds that two packets in her sample contain a free gift.

iii State, with a reason, the conclusion she should draw. **[2]**

*Cambridge International AS & A Level Mathematics 9709 Paper 72 Q4 November 2012*

10 An engineering test consists of 100 multiple-choice questions. Each question has five suggested answers, only one of which is correct. Ashok knows nothing about engineering, but he claims that his general knowledge enables him to get more questions correct than just by guessing. Ashok actually gets 27 answers correct. Use a suitable approximating distribution to test at the 5% significance level whether his claim is justified. **[5]**



*Cambridge International AS & A Level Mathematics 9709 Paper 71 Q2 November 2011*

11 Jeevan thinks that a six-sided die is biased in favour of six. To test this, Jeevan throws the die 10 times. If the die shows a six on at least 4 throws out of 10, she will conclude that she is correct.

i State appropriate null and alternative hypotheses. **[1]**

ii Calculate the probability of a Type I error. **[3]**

iii Explain what is meant by a Type II error in this situation. **[1]**

iv If the die is actually biased so that the probability of throwing a six is $\frac{1}{2}$, calculate the probability of a Type II error. **[3]**

*Cambridge International AS & A Level Mathematics 9709 Paper 72 Q6 June 2011*

12 In Europe the diameters of women's rings have mean 18.5 mm. Researchers claim that women in Jakarta have smaller fingers than women in Europe. The researchers took a random sample of 20 women in Jakarta and measured the diameters of their rings. The mean diameter was found to be 18.1 mm. Assuming that the diameters of women's rings in Jakarta have a normal distribution with standard deviation 1.1 mm, carry out a hypothesis test at the 2.5% level to determine whether the researchers' claim is justified. **[5]**

*Cambridge International AS & A Level Mathematics 9709 Paper 71 Q1 June 2009*

# Chapter 2
# The Poisson distribution

**In this chapter you will learn how to:**

- understand the Poisson distribution as a probability model
- calculate probabilities using the Poisson distribution
- solve problems involving linear combinations of independent Poisson distributions
- use the Poisson distribution as an approximation to the binomial distribution
- use the normal distribution as an approximation to the Poisson distribution
- carry out hypothesis testing of a Poisson model.

**PREREQUISITE KNOWLEDGE**

| Where it comes from | What you should be able to do | Check your skills |
|---|---|---|
| Probability & Statistics 1, Chapter 7 | Calculate probabilities using the binomial distribution. | **1** If $X \sim B(10, 0.2)$, find: **a** $P(X = 7)$ **b** $P(X < 2)$ |
| Probability & Statistics 1, Chapter 8 | Use normal distribution tables. | **2** If $X \sim N(16, 1.8)$, find: **a** $P(X < 20)$ **b** $P(13 < X < 15)$ |
| Pure Mathematics 2 & 3, Chapter 2 | Evaluate exponential expressions. | **3** Giving your answers to 3 significant figures, evaluate: **a** $e^{-2}$ **b** $e^{-3.6}$ **c** $e^{-9}$ |
| Chapter 1 | Formulate and carry out hypothesis testing. | **4** State the null and alternative hypotheses and test statistic for a test for the population mean for the random variable $X \sim N(42, 8)$, sample value 45; two-tailed test at 10% level of significance. |



## Why do we study the Poisson distribution?

In Probability & Statistics 1, you learnt about discrete probability distributions. In the binomial distribution there are a fixed number of trials with only two possible outcomes, usually designated 'success' or 'failure', for each trial, and you count the number of successes. In the geometric distribution there are, again, two possible outcomes, success or failure, and you count the number of trials up to and including the one in which the first success is obtained.

The following situation is an example where you count occurrences of an outcome, but the situation does not fit either the binomial or geometric distribution model. In 1898, Ladislaus Bortkiewicz investigated the number of soldiers in the Prussian army accidentally killed by horse kicks. It was rare for soldiers to be accidently killed in this way. There was no way to predict when it might happen: horse kicks happen at random. Bortkiewicz collected the following data over 20 years for 14 army corps.

| Number of deaths | 0 | 1 | 2 | 3 | 4 |
|---|---|---|---|---|---|
| Frequency | 144 | 91 | 32 | 11 | 2 |

Each data point in the table represents the number of accidental deaths from horse kicks in one corps in one year; so there are $144 + 91 + 32 + 11 + 2 = 280$ data points in total.

From the data, the mean average number of accidental deaths from horse kicks in one year is 0.7 (as shown in Section 2.1). The modal number is zero. The most recorded was four. However, it was possible, although unlikely, for the number of accidental deaths from horse kicks in an army corps to be much greater than four, and it was not possible to predict when during each year an accidental death from a horse kick would happen. In this situation, you count the number of outcomes, accidental deaths from horse kicks, over a fixed interval, one year.

Numerous other situations have the same characteristics of rare outcomes occurring singly and at random within a fixed interval. For example, the number of eagles nesting in a region or the number of bubbles in an item of glassware or the number of typing errors per page made by a good typist or the number of an insect type in a square metre of farmland. Notice, too, that the fixed interval can be a period of time or space.

Consider this example. A clinic treats people with severe insect bites. It keeps records for 100 days. On most days, no more than three people with severe insect bites attend the clinic, but on one day there are 16 people. The bar chart shows the frequencies.

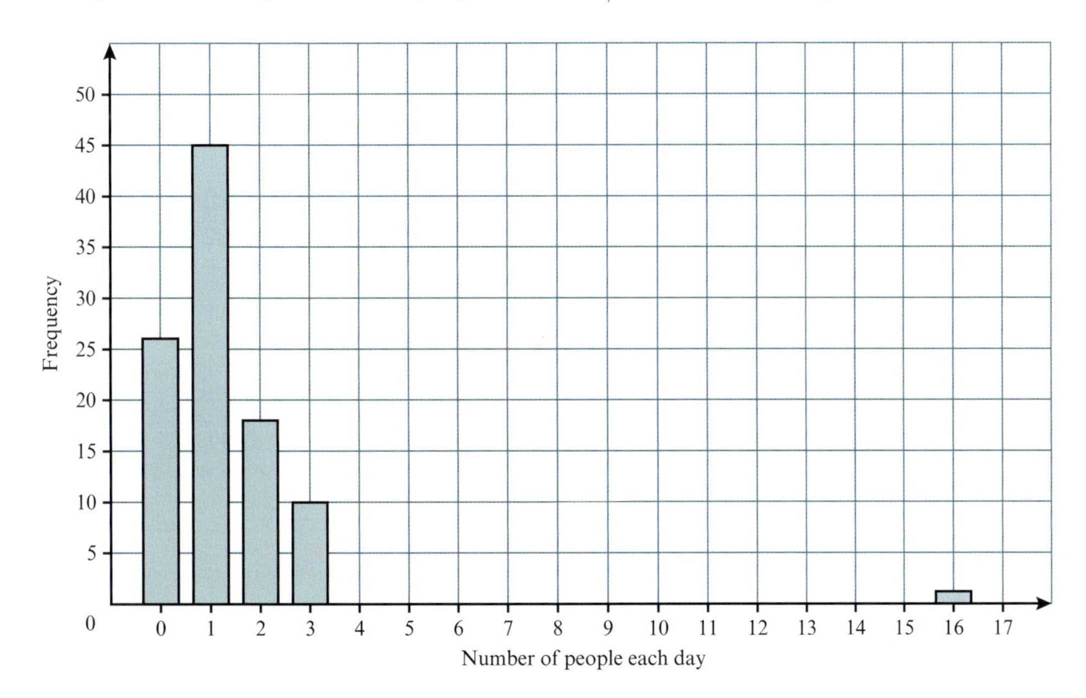


The bar chart shows what we may expect if each member of a population has the same low probability of attending the clinic for a severe insect bite. In theory, the maximum number of people attending the clinic is infinite, but large numbers, much bigger than the mean, are not expected.

To model these situations, you need a different discrete probability distribution from those you have already studied. This type of discrete distribution is a **Poisson distribution**.

## 2.1 Introduction to the Poisson distribution

First, let us explore Bortkiewicz's data further.

| **Number of deaths, $x$** | 0 | 1 | 2 | 3 | 4 |
|---|---|---|---|---|---|
| **Frequency, $f$** | 144 | 91 | 32 | 11 | 2 |

Here are calculations for the mean and the variance.

$$\bar{x} = \frac{\Sigma fx}{\Sigma f} = \frac{(0 \times 144) + (1 \times 91) + (2 \times 32) + (3 \times 11) + (4 \times 2)}{144 + 91 + 32 + 11 + 2} = \frac{196}{280} = 0.7$$

$$\text{Var}(x) = \frac{\Sigma fx^2}{\Sigma f} - \bar{x}^2 = \frac{(0^2 \times 144) + (1^2 \times 91) + (2^2 \times 32) + (3^2 \times 11) + (4^2 \times 2)}{144 + 91 + 32 + 11 + 2} - 0.7^2$$

$$= \frac{350}{280} - 0.7^2 = 0.76$$

We can see that the mean and variance are quite close in value. When working with experimental data where the outcome or event occurs at random in a given interval and the mean and variance are almost the same value, a Poisson distribution is a good choice to model the data further. For reasons we will not explore here, when working with experimental data we use the mean rather than the variance as the parameter to describe a Poisson distribution.

**KEY POINT 2.1**

A Poisson distribution can be used to model a discrete probability distribution in which the events occur singly, at random and independently, in a given interval of space or time. The mean and variance of a Poisson distribution are equal; hence, a Poisson distribution has only one parameter.

Let us explore another example.

To take away some of the pressure from an accident and emergency hospital department during weekdays, a minor injuries clinic opens. To determine possible staffing requirements, the following data for 250 30-minute intervals are collected.

| Number of patients arriving per 30 minutes, $x$ | 0 | 1 | 2 | 3 | 4 | 5 | $>5$ |
|---|---|---|---|---|---|---|---|
| Frequency, $f$ | 45 | 81 | 58 | 40 | 22 | 4 | 0 |

For these data, the mean number of patients per 30-minute interval is $\frac{\Sigma fx}{\Sigma f} = \frac{425}{250} = 1.7$ and the variance is $\frac{\Sigma fx^2}{\Sigma f} - \bar{x}^2 = \frac{1125}{250} - 1.7^2 = 1.61$.

A statistician notes that patients arriving at the clinic generally do so independently of each other at random intervals. The data are collected for fixed 30-minute intervals. The mean and variance, 1.7 and 1.61, respectively, are quite close in value. All these factors suggest the Poisson distribution is a suitable model to use. The Poisson model for the data is as follows:

If $X$ is the random variable 'number of patients arriving in 30 minutes', then $X \sim \text{Po}(1.7)$, $\text{E}(X) = 1.7$ and $\text{Var}(X) = 1.7$.

For a Poisson distribution with mean $\lambda$ the probability of an event $r$ occurring is given as:

$$\text{P}(X = r) = \text{e}^{-\lambda} \frac{\lambda^r}{r!}.$$

**FAST FORWARD**

A proof of this formula is not needed at this stage, but it will be explained later in the chapter (Section 2.3); for now we will just learn how to use it.

**REWIND**

You may have already met the mathematical constant e in the Pure Mathematics 2 & 3 Coursebook. If you have not, then you need to know that $\text{e} = 2.7183$ (to 4 decimal places) and you can calculate powers of e using a calculator.

**DID YOU KNOW?**

The mathematical constant e is an irrational number that is the base of natural logarithms. Leonard Euler famously discovered the value when solving a problem posed by another mathematician, Jacob Bernoulli (pictured), to find $\lim_{n\to\infty}\left(1+\frac{1}{n}\right)^n$.

**WEB LINK**

You can watch a more detailed explanation in the *e (Euler's Number)* numberphile clip on YouTube.

For the minor injuries clinic example, if we have $r$ patients arriving in 30-minute intervals, then $\text{P}(X = r) = \text{e}^{-1.7} \frac{1.7^r}{r!}$.

This is the graph of the probabilities of numbers of patients arriving in 30-minute intervals.

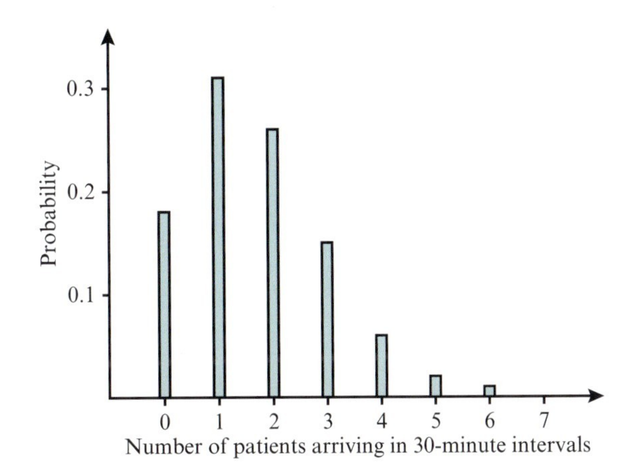


We can see from the graph that the probabilities peak at lower values of $X$ and begin to get much smaller for values of $X$ greater than 4. It is possible, although unlikely, that a large number of patients will arrive in a randomly chosen 30-minute time interval.

Using the probability formula for a Poisson distribution with mean = 1.7, $P(X = r) = e^{-1.7} \frac{1.7^r}{r!}$, we can work out expected frequencies for different numbers of patients. For example, three patients arriving in a 30-minute interval $P(X = 3) = e^{-1.7} \frac{1.7^3}{3!} = 0.150$. The expected number of patients is $250 \times P(X = 3) = 250 \times 0.150 = 37$, to the nearest integer. The following table shows the probabilities and the expected frequencies, together with the observed frequencies of patients per 30 minutes from the previous table.

| Number of patients per half-hour ($r$) | 0 | 1 | 2 | 3 | 4 | 5 | 6 | 7 |
|---|---|---|---|---|---|---|---|---|
| $P(X = r) = e^{-1.7} \frac{1.7^r}{r!}$ | 0.183 | 0.311 | 0.264 | 0.150 | 0.0636 | 0.0216 | 0.00612 | 0.00149 |
| Expected frequency $250 \times P(X = r)$ | 46 | 78 | 66 | 37 | 16 | 5 | 2 | 0 |
| Observed frequency | 45 | 81 | 58 | 40 | 22 | 4 | 0 | 0 |

A comparison of the expected frequencies with the observed frequencies shows that the numbers are all reasonably close. This allows us to use the Poisson distribution as a model.

Using the data, the statistician makes calculations and advises the clinic that there is about a 3% chance, $\frac{5 + 2 + 0}{250} \times 100$, of staff being needed to deal with more than four patients every 30 minutes.

**EXPLORE 2.1**

The data collected by Bortkiewicz given in the introduction is one of the earliest examples of a Poisson distribution being used as a model. For Bortkiewicz's data, use the formula for Poisson probabilities and the calculated mean, 0.7, to work out the expected frequencies. Compare your calculated expected frequencies with the actual data. Does the Poisson distribution appear to be a reasonable model for these data? What predictions can you make from your theoretical model of the data?

**KEY POINT 2.2**

When modelling data using a Poisson distribution:

- Work out the mean and variance and check if they are approximately equal.

  If the mean and variance are not approximately equal, the Poisson distribution is not a suitable model to use with the data.
- Use the mean to calculate probabilities and expected frequencies.
- Compare expected frequencies with observed frequencies.

**KEY POINT 2.3**

If the random variable $X$ has a Poisson distribution with parameter $\lambda$, where $\lambda > 0$, we write $X \sim \text{Po}(\lambda)$ and:

- $\text{P}(X = r) = e^{-\lambda}\dfrac{\lambda^r}{r!}$, where $r = 0, 1, 2, \ldots$
- $\text{E}(X) = \lambda$
- $\text{Var}(X) = \lambda$



**EXPLORE 2.2**

Using graphing software, such as GeoGebra, explore the shape of the Poisson distribution for different values of $\lambda$.

- What features do you notice?
- How does the shape of the graph change as the value of $\lambda$ varies?
- For different values of $\lambda$, find the value of $r$ where $\text{P}(X = r)$ is at its maximum.
- What do you notice?
- Can you give a general answer in terms of $\lambda$?

**WORKED EXAMPLE 2.1**

The number of calls to a consumer hotline is modelled by a Poisson distribution with a mean call rate of six per minute. Calculate the probability that in a given minute there will be:

**a** nine calls

**b** three or fewer calls

**c** more than one call

**d** at least one call.

**Answer**

$X \sim \text{Po}(6)$ Always state the random variable you are working with.

a $P(X=9) = e^{-6}\dfrac{6^9}{9!} = 0.0688$ Apply the formula.

b $P(X \leqslant 3) = e^{-6}\dfrac{6^0}{0!} + e^{-6}\dfrac{6^1}{1!} + e^{-6}\dfrac{6^2}{2!} + e^{-6}\dfrac{6^3}{3!}$ Note that 'three or fewer' includes three.

$= e^{-6}\left(1 + 6 + \dfrac{6^2}{2} + \dfrac{6^3}{3!}\right)$ Factorise out $e^{-6}$ to make the working more straightforward.

$= 0.151$

c $P(X > 1) = 1 - \left(e^{-6}\dfrac{6^0}{0!} + e^{-6}\dfrac{6^1}{1!}\right) = 1 - 0.0174$ Remember that the probability 'greater than' = 1 – probability 'less than or equal to'.

$= 0.983$

d $P(X \geqslant 1) = 1 - P(X=0) = 1 - e^{-6}\dfrac{6^0}{0!}$ Remember that $P(X \geqslant 1) = P(X > 0)$.

$= 1 - 0.00248 = 0.998$

**TIP**

When working out more than one Poisson probability, first factorise out the term $e^{-\lambda}$.



**WORKED EXAMPLE 2.2**

A typesetter makes, on average, five errors per page of typing.

a State the assumptions made to model the average number of errors per page as a Poisson distribution.

b In a book with 200 pages, on how many pages would you expect to find, at most, two errors?

**Answer**

a Assumptions: errors occur independently, singly and at random. Clearly state your assumptions.

b $X \sim \text{Po}(5)$ First state the random variable. Remember that 'at most' means up to and including that value.

$P(X \leqslant 2) = e^{-5}\left(1 + 5 + \dfrac{5^2}{2!}\right) = 0.124652019...$

$200 \times 0.124652019... = 24.9$ or 25 pages

## EXERCISE 2A

1 $X \sim \text{Po}(4)$. Calculate:

a $\text{P}(X = 5)$ b $\text{P}(X < 3)$ c $\text{P}(X > 0)$

2 $X \sim \text{Po}(7)$. Calculate:

a $\text{P}(X = 6)$ b $\text{P}(X \geqslant 3)$

3 $X \sim \text{Po}(1.5)$. Calculate:

a $\text{P}(X = 4)$ b $\text{P}(X \leqslant 2)$ c $\text{P}(X > 2)$

M 4 In a particular town, it was found that potholes occur independently and at random at a rate of five in a 1 kilometre stretch of road.

a What is the probability that in a randomly chosen 1 kilometre stretch of road in this town there will be fewer than three potholes?

b Explain why this model may not be applicable to other towns.

M 5 A fire station receives, on average, one emergency call per hour.

a i What assumptions do you need to make to model this situation by a Poisson distribution?

ii Are the assumptions reasonable in this situation?

b Assuming the assumptions are reasonable, what is the probability of four calls being received between 11 pm and midnight?



M 6 A manufacturer of chocolate bars states that the number of whole hazelnuts in a randomly chosen 100 g hazelnut chocolate bar can be modelled as a random variable having a Poisson distribution with mean 7.2.

a Find the probability that in a randomly chosen 100 g hazelnut chocolate bar there are:

i exactly eight whole hazelnuts

ii at least four whole hazelnuts.

b Describe how you could check the manufacturer's statement.

M 7 A handful of rice grains is scattered at random onto a chessboard. Jo counts the number of rice grains in each of the 64 squares on the chessboard.

| Number of rice grains per square | 0 | 1 | 2 | 3 | 4 | 5 | 6 | $>6$ |
|---|---|---|---|---|---|---|---|---|
| Number of squares | 12 | 20 | 20 | 7 | 3 | 1 | 1 | 0 |

a Use appropriate calculations to show that it may be possible to model the distribution of rice grains using a Poisson distribution.

b Using an appropriate value for the parameter, find the expected distribution of numbers of rice grains in the squares.

**EXPLORE 2.3**

For the situation in Exercise 2A Question 7 to follow a Poisson distribution, the handful of rice is scattered at random onto a chessboard. Will similar results be true for different-sized boards of squares? Or for a board of triangles?

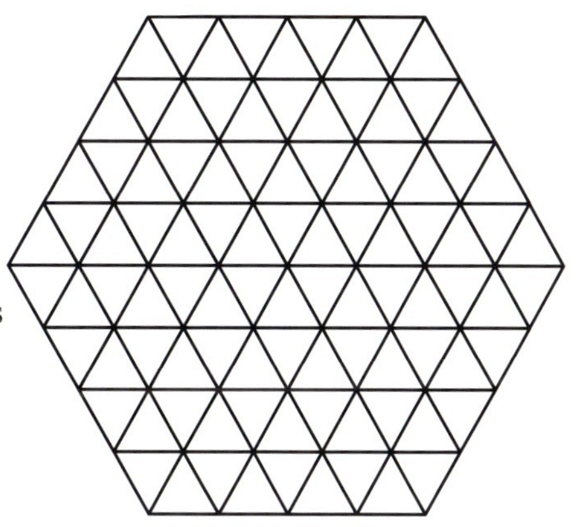

Experiment for yourself. Tabulate your results and calculate the mean and variance per square or triangle.

What do your results suggest?

## 2.2 Adapting the Poisson distribution for different intervals

In the minor injuries clinic example, we chose to collect data on the number of patients arriving at the clinic in 30-minute intervals. We could have chosen to collect the data in 60-minute intervals or 10-minute intervals. Suppose we had collected the data across the whole time interval of 7500 ($250 \times 30$) minutes. We would still have noted 425 patients arriving. The mean arrival rate of patients would be 1.7 per 30 minutes. We could also calculate the average rate of patients arriving per 60 minutes; $425 \times \frac{60}{7500} = 3.4$ patients per 60 minutes. Or the average rate per 10 minutes; $425 \times \frac{10}{7500} = \frac{17}{30}$ patients per 10 minutes. The key point here is that the Poisson distribution is based on an average rate, *something per something*. If we know the rate for a specific interval then we can adapt the parameter to use for multiples of that interval.



**KEY POINT 2.4**

In a Poisson distribution, events occur at a constant rate; the mean average number of events in a given interval is proportional to that interval.

#### WORKED EXAMPLE 2.3

People arrive at random and independently at a post office at an average rate of two people every 5 minutes. Work out the probability of:

**a** three people arriving in a 10-minute period

**b** more than four people arriving in a half-hour period

**c** five people arriving in a 4-minute period

**d** one person arriving in a 1-minute period.

**Answer**

**a** $X \sim \text{Po}(4)$

$P(X = 3) = e^{-4}\frac{4^3}{3!} = 0.195$

First find the value of $\lambda$; 10 minutes is double 5 minutes so $\lambda$ is $2 \times 2$.

**b** $X \sim \text{Po}(12)$

$P(X > 4) = 1 - P(X \leqslant 4)$

$= 1 - e^{-12}\left(1 + 12 + \frac{12^2}{2!} + \frac{12^3}{3!} + \frac{12^4}{4!}\right)$

$= 1 - 0.0076 = 0.9924$ or $0.992$

First find the value of $\lambda$; half-hour $= 6 \times 5$ minutes so $\lambda$ is $6 \times 2$.

Note that $> 4$ doesn't include the value 4.

**c** $X \sim \text{Po}(1.6)$

$P(X = 5) = e^{-1.6}\frac{1.6^5}{5!} = 0.0176$

$\lambda$ does not need to be either an exact multiple or an integer; here $\lambda = \frac{4}{5} \times 2 = 1.6$.

**d** $X \sim \text{Po}(0.4)$

$P(X = 1) = e^{-0.4}\frac{0.4^1}{1!}$

$= 0.268$

First find the value of $\lambda$; 1 minute is one-fifth of 5 minutes so $\lambda = \frac{1}{5} \times 2 = 0.4$.

The time period can be smaller than the original period given.

Remember the mean does not need to be a whole number.



#### WORKED EXAMPLE 2.4

The number of breakages at a restaurant in a randomly chosen week can be modelled as a random variable having a Poisson distribution with mean 0.8.

**a** Work out the probability of the following.

**i** Exactly one breakage in 1 week.

**ii** Exactly one breakage in a randomly chosen 3-week period.

**b** The manager offers staff a bonus if there are no breakages in 6 consecutive weeks. What is the probability that staff receive a bonus?

**Answer**

**a i** $X \sim \text{Po}(0.8)$

$P(X = 1) = e^{-0.8}\frac{0.8^1}{1!} = 0.359$

Use the probability formula directly.

ii $X \sim \text{Po}(2.4)$

$\text{P}(X = 1) = e^{-2.4}\,\frac{2.4^1}{1!} = 0.218$

A randomly chosen 3-week period is a multiple of the interval.

b $X \sim \text{Po}(0.8)$

$\text{P}(X = 0) = \left(e^{-0.8}\,\frac{0.8^0}{0!}\right)^6 = 0.00823$

Or $X \sim \text{Po}(4.8)$

$\text{P}(X = 0) = e^{-4.8}\,\frac{4.8^0}{0!} = 0.00823$

The two approaches highlight why you can use multiples of an interval.

Calculate the probability of no breakages in 1 week then raise to power 6 for 6 weeks.

Or find the average rate for 6 weeks, $6 \times 0.8$, and use this average for a single 6-week interval.

## EXERCISE 2B

1 Potholes occur independently and at random at a rate of five in a stretch of road 1 kilometre long. Calculate the probability that in a randomly chosen stretch of road 2 kilometres long there will be:

a exactly eight potholes

b fewer than two potholes.

2 The number of faults in a roll of wallpaper can be modelled as a random variable having a Poisson distribution with mean 0.6. Find the probability that a decorator using four rolls of wallpaper for a room finds no faults in the wallpaper.

M 3 The number of flaws in a given length of cloth occur at the rate of 1.6 per metre. State the assumptions you need to make to model this situation as a Poisson distribution. Find the probability that:

a in a 5-metre length of cloth there are no flaws

b in a $\frac{1}{2}$-metre length of cloth there are two or more flaws.

M 4 The number of cars passing a point on a road can be modelled as a random variable having a Poisson distribution with mean two cars per 5 minutes.

a What is the probability in a randomly chosen 20-minute period that more than three cars will pass that point on the road?

b What conclusions might you draw if no cars pass that point in the randomly chosen 20-minute period?

c How might installing traffic lights at one end of the road affect the Poisson model?

5 Over a long period of time, a plumber finds that, on average, he receives two emergency calls per week. Work out the probability of:

a no emergency calls in a two-week period

b one emergency call on one day (assume the plumber is available for emergency calls five days per week).

6 A typist makes, on average, one error for every 200 keyboard strokes. Assuming the errors occur independently and at random, find the probability that:

a in a document requiring 400 keyboard strokes there is, at most, one error

b in two documents requiring 400 keyboard strokes there is, at most, one error.

M 7 The number of orders placed at an online store is 4500 per hour.

a What assumptions do you need to make to model the number of orders placed at an online store using a Poisson distribution?

**b** Assuming the number of orders placed can be modelled as a Poisson distribution, find the probability of:

**i** zero orders occurring per second **ii** one order occurring per second.

**8** Over a period of time, a consumer complaints department notes that it receives an average of 4.2 emails per hour.

**a** What assumptions do you need to make to model the number of email complaints received by a consumer complaints department as a Poisson distribution?

**b** Why might it be unlikely that all complaints are made independently of each other?

**c** Are complaints more likely to be made at certain times rather than others?

**d** Assuming the number of emails received can be modelled as a Poisson distribution, find the probability that the consumer complaints department receives:

**i** nine emails in 2 hours **ii** three emails per minute.

**DID YOU KNOW?**

In 1837 the French mathematician Simeon-Denis Poisson published his probability theory in his research work on the probability of judgements in criminal and civil matters, in which he theorised on the number of wrongful convictions. The Poisson distribution is named in his honour.



## 2.3 The Poisson distribution as an approximation to the binomial distribution

A manufacturer makes plastic pipe in a continuous length before cutting the pipe into shorter lengths to sell. Suppose there are, on average, four defects per metre of pipe; then we have a Poisson distribution Po(4).

On average, how many defects would there be in a 10 cm length or in a 1 cm length or in a 1 mm length of plastic pipe? It would be reasonable to say that for some length of pipe there will be either zero or one defect, where the probability of more than one defect is so small that it may be ignored.

If we now use $n$ pieces of this smaller length of pipe to make a 1 m length of pipe, the probability of one defect in a length of pipe is $\frac{4}{n}$. We have a fixed number of pieces with only two outcomes, zero or one defect, and hence a binomial distribution $\text{B}\left(n, \frac{4}{n}\right)$.

From this example, we can see that the Poisson distribution and the binomial distribution are related. Our question is for what values of $n$ and $p$ is it reasonable to use a Poisson distribution to approximate to the binomial distribution?

In this example, we have $\text{B}\left(n, \frac{4}{n}\right) \approx \text{Po}(4)$.

For the binomial, mean $= n \times \frac{4}{n} = 4$, which is the same as the mean value for Po(4).

The variance of the binomial $= n \times \frac{4}{n}\left(1 - \frac{4}{n}\right)$. As $n$ gets larger, $\frac{4}{n}$ gets very small and $1 - \frac{4}{n} \approx 1$. And so the variance $= 4$, which is the same as the variance for Po(4). This shows that the approximation of a binomial distribution by a Poisson distribution improves as $n$ becomes larger.

Let us explore some graphs of these **discrete random variables**.

These graphs show Poisson distributions for different mean values.

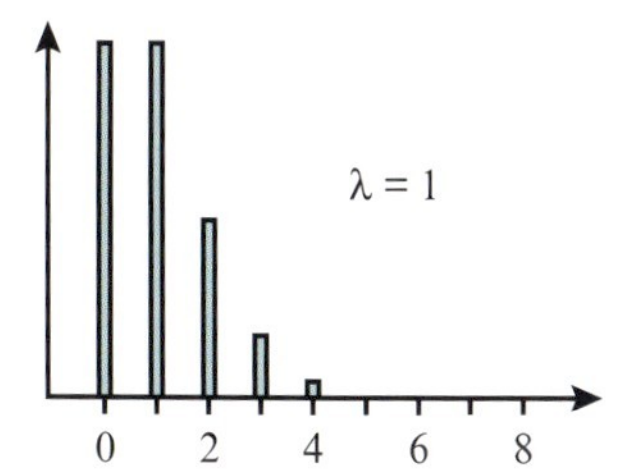


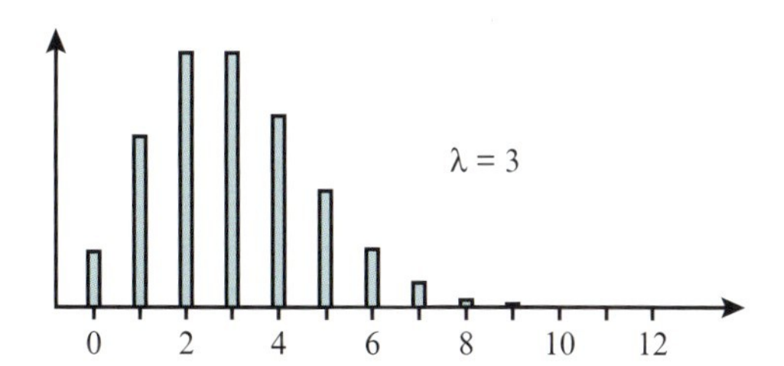


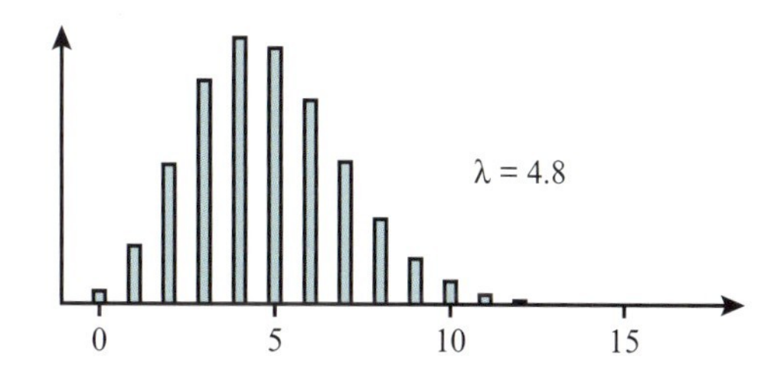


Graphs of Poisson distributions are always skewed, but when the mean is small, a graph of the Poisson distribution is very skewed. As the mean increases, the graph of a Poisson distribution becomes more symmetrical.

These graphs show binomial distributions for different values of $n$ and $p$.

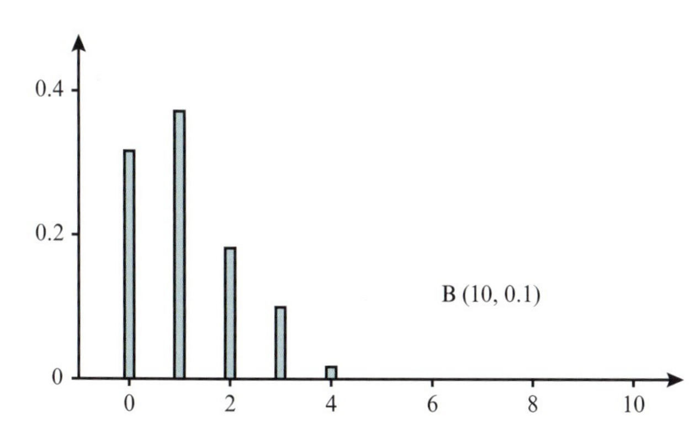


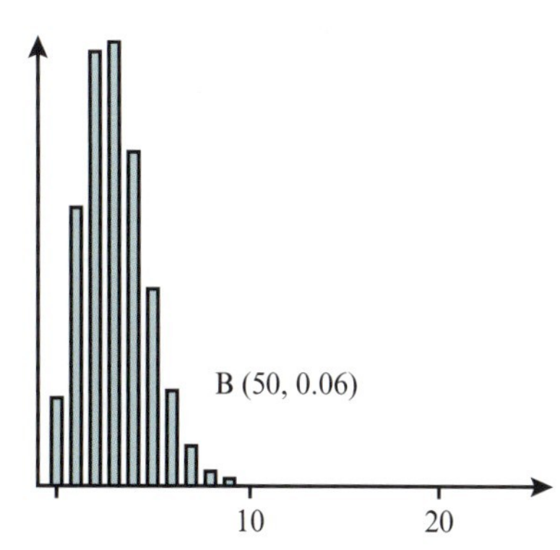


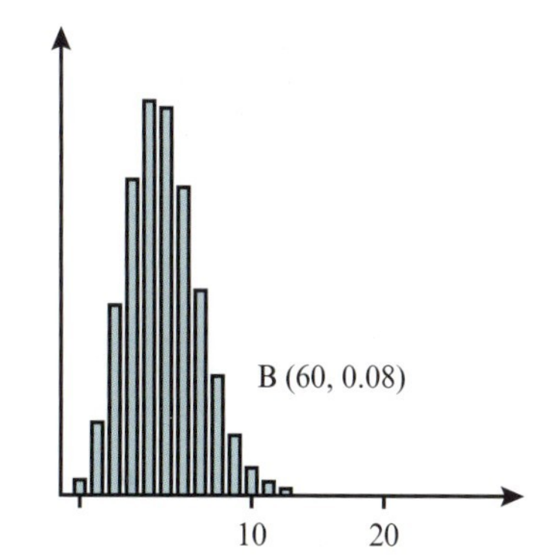


For a binomial distribution $\text{E}(X) = np$, and for a Poisson distribution $\text{E}(X) = \lambda$.

Comparing the graphs of the two distributions with the same mean value, $np = \lambda$, we see that for B(20, 0.05 and Po(1) the Poisson graph is very skewed whereas the binomial graph is more symmetrical.

As $n$ becomes large and $p$ becomes small, we can see the graph for B(60, 0.08) becoming more similar to the graph for Po(4.8).

**KEY POINT 2.5**

For a binomial distribution when the value of $n$ is large and $p$ is small (implying occurrence of the event is rare), such that $np$ is moderate (as a guide, $n > 50$ and $np < 5$), the Poisson distribution with mean $np$ can be used as an approximation for the binomial distribution.

**EXPLORE 2.4**

Many binomial distributions give rise to the same Poisson distribution. For example, Po(4.8) could be approximated from B(100, 0.048) or B(10, 0.48) or B(20, 0.24). Not all of these are equally good approximations. Use graphing software, such as GeoGebra, to explore different graphs of binomial and Poisson distributions where $np = \lambda$.

To find out how good an approximation a Poisson distribution is to the binomial distribution, let's compare exact probabilities and approximate probabilities.

**REWIND**

Recall from Probability & Statistics 1 Coursebook, Chapter 7, that this situation can be modelled by a binomial situation with $n = 375$ and $p = 0.012$.

Gareth makes glassware. An item of glassware is imperfect if it contains bubbles. Over the years, he finds that the probability of any item containing bubbles is 0.012. Gareth made a batch of 375 items of glassware.

Let $X$ be the random variable 'number of glassware items containing bubbles'. Then $X \sim \text{B}(375, 0.012)$.

Assuming the presence of a bubble in one item of glassware is independent of the presence of a bubble in another item of glassware, then $np = 375 \times 0.012 = 4.5$ and $\text{B}(375, 0.012) \approx \text{Po}(4.5)$.

The following table shows some probabilities of finding bubbles in an item of glassware.

| Number of bubbles, $r$ | $\text{P}(X = r)$ | 0 | 1 | 2 | 3 |
|---|---|---|---|---|---|
| **Binomial probability** | $\binom{375}{r} 0.012^r (1 - 0.012)^{375-r}$ | 0.0108 | 0.0492 | 0.112 | 0.169 |
| **Poisson approximation** | $\frac{e^{-4.5} 4.5^r}{r!}$ | 0.0111 | 0.0450 | 0.112 | 0.169 |

We can see that the probabilities, correct to 3 significant figures, are almost the same for all these values; hence, the Poisson is a good approximation to the binomial to use in this situation.



## WORKED EXAMPLE 2.5

A company produces electrical components. Past records show that the proportion of faulty components is 0.4%. Hari buys a box of 1000 electrical components. Using a suitable approximation, work out the probability that more than five components are faulty.

**Answer**

Let the random variable $X$ be 'the number of faulty components'. Then $X \sim \text{B}(1000, 0.004)$.

The situation is binomial.
Probability $0.4\% = \frac{0.4}{100} = 0.004$

$\text{B}(1000, 0.004) \approx \text{Po}(4)$

Poisson approximation is suitable as $n$ is large, $p$ is small and $np = 1000 \times 0.004 = 4$.

$\text{P}(X > 5) = 1 - \text{P}(X \leqslant 5)$

$= 1 - e^{-4}\left(\frac{4^0}{0!} + \frac{4^1}{1!} + \frac{4^2}{2!} + \frac{4^3}{3!} + \frac{4^4}{4!} + \frac{4^5}{5!}\right)$

Factorise out to simplify the working.

$= 1 - 0.7851$

$= 0.2150$

## EXERCISE 2C

1 For the random variable $X$, where $X \sim \text{B}(60, 0.05)$, use a suitable approximation to find:

a $\text{P}(X < 4)$ b $\text{P}(X \geqslant 4)$

2 For the random variable $X$, where $X \sim \text{B}(120, 0.02)$, use a suitable approximation to find:

a $\text{P}(X \leqslant 5)$ b $\text{P}(X > 2)$

3 For the random variable $X$, where $X \sim B(200, 0.01)$, use a suitable approximation to find $P(6 < X < 10)$.

PS 4 Past records show that the proportion of faulty resistors manufactured by a company is 0.2%. Robert buys a box of 450 resistors. Using a suitable approximation, work out the probability that fewer than three resistors are faulty.

PS 5 A machine is known to produce defective components, at random and independently of each other, on average 0.24% of the time. In a production of 500 components, state a suitable approximating distribution and calculate the probability that, at most, three components will be defective.

M 6 A rare reaction to a prescribed medicine occurs in 0.05% of patients.

a Using a suitable approximation, find the probability that in a random sample of 3000 patients prescribed this medicine, four or more will suffer the rare reaction.

b In a random group of $n$ patients, the probability that none suffer the rare reaction is 0.001. Work out the value of $n$.

PS 7 For a certain flower, a seed mutation occurs at random with probability 0.0004. A total of 12 000 seeds germinate. Let $X$ be the number of seeds that germinate and carry the mutation.

a Justify using the Poisson distribution as an approximating distribution for $X$.

b Use your approximating distribution to find $P(X \leqslant 3)$.

c Calculate $P(X \leqslant 3)$ given that $X > 1$.



Consider again the example of the manufacturer of plastic pipe at the start of Section 2.3. In that example there were $n$ small pieces of pipe in which there was only zero or one defect in each piece, and hence $X \sim B\left(n, \frac{4}{n}\right)$. To ensure each piece does not contain more than one defect, the pieces would need to be very small and we would have a large number of pieces.

Suppose $n = 1000$, then $P(X = 0) = \binom{1000}{0}\left(\frac{4}{1000}\right)^0\left(1 - \frac{4}{1000}\right)^{1000} = 0.018169\ldots$

For $n = 10\,000$, $P(X = 0) = \binom{10\,000}{0}\left(\frac{4}{10\,000}\right)^0\left(1 - \frac{4}{10\,000}\right)^{10\,000} = 0.018300\ldots$

For $n = 100\,000$, $P(X = 0) = \binom{100\,000}{0}\left(\frac{4}{100\,000}\right)^0\left(1 - \frac{4}{100\,000}\right)^{100\,000} = 0.018315\ldots$

We can see that for increasing values of $n$ the probabilities tend towards $0.018315\ldots = e^{-4}$. This is the value given by the Poisson probability for $P(X = 0)$, where $X \sim Po(4)$.

Let's move on to explore the probabilities for one defect using the binomial $X \sim B\left(n, \frac{4}{n}\right)$.

For $n = 1000$, $P(X = 1) = \binom{1000}{1}\left(\frac{4}{1000}\right)^1\left(1 - \frac{4}{1000}\right)^{999} = 0.072969\ldots$

For $n = 10\,000$, $P(X = 1) = \binom{10\,000}{1}\left(\frac{4}{10\,000}\right)^1\left(1 - \frac{4}{1000}\right)^{9999} = 0.072323\ldots$

If we continue to calculate probabilities for one defect, increasing the values of $n$, the result will tend towards 0.073262…, and this is the value given by the Poisson probability for $P(X = 1)$ for $e^{-4}$.

In general, for $X \sim \text{B}\left(n, \dfrac{\lambda}{n}\right)$ and $r$ defects:

$$\text{P}(X = r) = \binom{n}{r}\left(\frac{\lambda}{n}\right)^r\left(1 - \frac{\lambda}{n}\right)^{n-r} = \frac{n(n-1)(n-2)\ldots(n-r+1)}{r!} \times \frac{\lambda^r}{n^r}\left(1 - \frac{\lambda}{n}\right)^{n-r}$$

$$= \frac{\lambda^r}{r!} \times \frac{n}{n} \times \frac{n-1}{n} \times \frac{n-2}{n} \times \ldots \times \frac{n-r+1}{n} \times \left(1 - \frac{\lambda}{n}\right)^{n-r}$$

As $n$ increases, the fractions $\dfrac{n-1}{n}$, $\dfrac{n-2}{n}$, ... tend towards 1; and $\left(1 - \dfrac{\lambda}{n}\right)^{n-r} \approx \left(1 - \dfrac{\lambda}{n}\right)^n$ since $r$ is negligible compared to $n$. We found earlier that $\left(1 - \dfrac{\lambda}{n}\right)^n$ tends towards $e^{-\lambda}$.

Hence, $\text{P}(X = r) = \dfrac{\lambda^r}{r!}e^{-\lambda}$, which is the probability formula for $X \sim \text{Po}(\lambda)$, the formula that was given in Section 2.1.

## 2.4 Using the normal distribution as an approximation to the Poisson distribution

In Explore 2.2 and in Section 2.3 we looked at graphs of Poisson distributions with different values of $\lambda$. This series of graphs shows the shape of the Poisson distribution as the value of $\lambda$ increases.

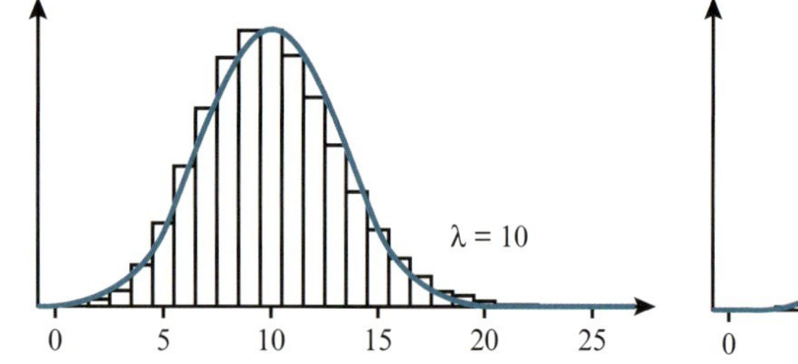


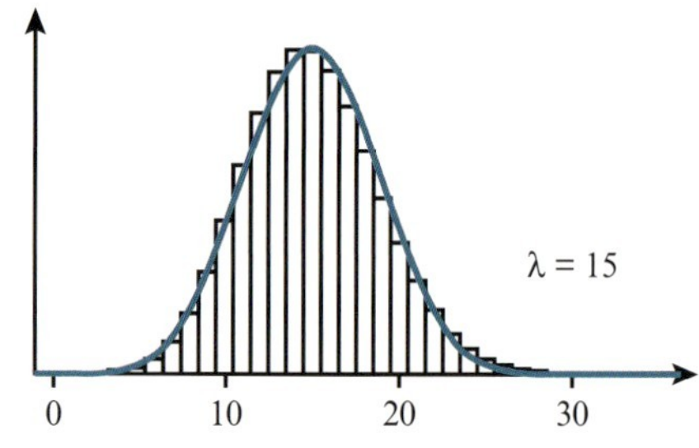


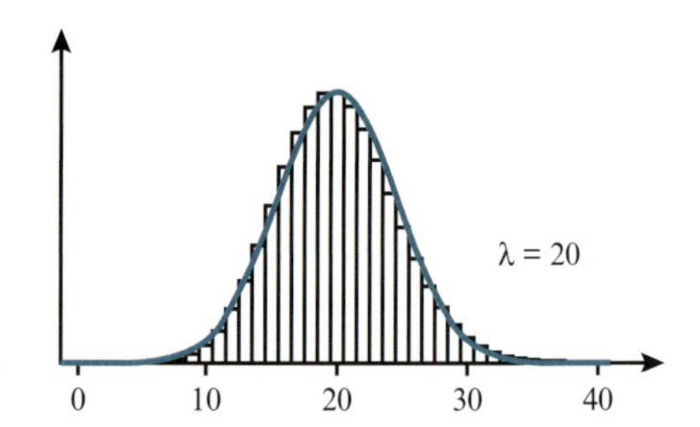




A normal distribution curve with the same mean and variance as the Poisson graph – that is, $\text{N}(\lambda, \lambda)$ – has been drawn on each graph.

Notice that as the value of $\lambda$ increases the Poisson graph improves as a fit to the normal curve. The shape of the Poisson distribution graph is always skewed; however, for larger values of $\lambda$, the Poisson distribution graph resembles the shape of a normal distribution.

The Poisson distribution is a limiting distribution of binomial distributions. A binomial distribution can be approximated by a normal distribution, so it seems reasonable to suppose that the Poisson distribution itself can be approximated by the normal distribution, with a continuity correction applied.

**REWIND**

In Probability & Statistics 1 Coursebook, Chapter 8, we saw that the normal distribution is a continuous probability distribution described by two parameters: mean and variance.

For a Poisson distribution the mean and variance are the same, and hence $\text{Po}(\lambda) \approx \text{N}(\lambda, \lambda)$.

A Poisson distribution is a discrete probability distribution and so to use the normal distribution as an approximation to the Poisson distribution a continuity correction has to be applied, just as one has to be applied when using the normal distribution as an approximation to a binomial distribution.

**REWIND**

In Probability & Statistics 1 Coursebook, Chapter 8, we saw that a continuity correction had to be applied when using the normal distribution as an approximation to a binomial distribution.

**KEY POINT 2.6**

For $\lambda > 15$, the Poisson distribution with mean $\lambda$ can be reasonably approximated by the normal distribution with mean $\lambda$ and variance $\lambda$, with a continuity correction applied. The accuracy of this approximation improves as $\lambda$ increases.

We can explore how this works, and how good an approximation it might be, by working out some probabilities. Assume that Manuela posts pictures on her social media page at random points in time. She posts, on average, 24 pictures each week.

Let the random variable $X$ be 'the number of pictures posted each week'.
Then $X \sim \text{Po}(24)$.

$$\text{P}(X = n) = \frac{e^{-24}24^n}{n!}$$

The table shows some probabilities for the number of pictures posted using the Poisson formula and using a normal approximation.

| Poisson | Normal approximation | Comments |
|---|---|---|
| $X \sim \text{Po}(24)$ | $Y \sim \text{N}(24, 24)$ | Apply the continuity correction when using the normal approximation. |
| $\text{P}(X = 30)$ $= \frac{e^{-24}24^{30}}{30!}$ $= 0.0363$ | $\text{P}(29.5 \leqslant Y \leqslant 30.5) = \text{P}\left(\frac{30.5-24}{\sqrt{24}} \leqslant Z \leqslant \frac{29.5-24}{\sqrt{24}}\right)$ $= \Phi(1.327) - \Phi(1.123)$ $= 0.9077 - 0.8692 = 0.0385$ | The approximate probability using the normal distribution is a range of values. |
| $\text{P}(29 < X < 31)$ $= \frac{e^{-24}24^{30}}{30!}$ $= 0.0363$ | $\text{P}(29.5 \leqslant Y \leqslant 30.5) = \text{P}\left(\frac{30.5-24}{\sqrt{24}} \leqslant Z \leqslant \frac{29.5-24}{\sqrt{24}}\right)$ $= \Phi(1.327) - \Phi(1.123)$ $= 0.9077 - 0.8692$ $= 0.0385$ | 29 and 31 are not included and so the range of values is 29.5 to 30.5. |
| $\text{P}(24 \leqslant X \leqslant 26)$ $= \frac{e^{-24}24^{24}}{24!}$ $+ \frac{e^{-24}24^{25}}{25!}$ $+ \frac{e^{-24}24^{26}}{26!}$ $= 0.231$ | $\text{P}(23.5 \leqslant Y \leqslant 26.5) = \text{P}\left(\frac{26.5-24}{\sqrt{24}} \leqslant Z \leqslant \frac{23.5-24}{\sqrt{24}}\right)$ $= \Phi(0.510) - \Phi(-0.102)$ $= 0.6950 - (1 - 0.5406)$ $= 0.236$ | 24 and 26 are included and so the range of values is 23.5 to 26.5. |

| Poisson | Normal approximation | Comments |
|---|---|---|
| $P(19 < X \leqslant 21)$ $= \frac{e^{-24}24^{20}}{20!} + \frac{e^{-24}24^{21}}{21!}$ $= 0.134$ | $P(19.5 \leqslant Y \leqslant 21.5) = P\left(\frac{21.5-24}{\sqrt{24}} \leqslant Z \leqslant \frac{19.5-24}{\sqrt{24}}\right)$ $= \Phi(-0.510) - \Phi(-0.919)$ $= (1-0.6950) - (1-0.8209)$ $= 0.126$ | 19 is not included but 21 is included, so the range of values is 19.5 to 21.5. |

From these calculations we can see that the probabilities using the normal distribution as an approximation to the Poisson distribution give the same set of values, correct to 2 decimal places, as the calculations using the Poisson probability formula in almost all the probabilities.

**WORKED EXAMPLE 2.6**

Over the years, a biologist notes that a species of turtle lays on average 60 eggs in each nest. The number of eggs laid in each nest follows a Poisson distribution, and is independent of the number of eggs laid in other nests. Calculate the probability that in a randomly chosen nest there are:

**a** exactly 50 eggs **b** over 74 eggs **c** 40 eggs or fewer.

**Answer**

Let $X$ be the random variable 'number of eggs in a nest'.
Then $X \sim \text{Po}(60) \approx \text{N}(60, 60)$.

Always state the distribution you are using and any approximating distribution.

**a** $P(X = 50) \approx P\left(\frac{50.5-60}{\sqrt{60}} \leqslant Z \leqslant \frac{49.5-60}{\sqrt{60}}\right)$
$= \Phi(-1.226) - \Phi(-1.356)$
$= (1 - 0.8899) - (1 - 0.9125) = 0.0226$

You could just use Poisson here; the question is included to highlight the process for a single value.

**b** $P(X > 74) = 1 - P(X < 75) \approx 1 - P\left(Z \leqslant \frac{74.5-60}{\sqrt{60}}\right)$
$= 1 - \Phi(1.872) = 1 - 0.9693 = 0.0307$

You could have written $1 - P(X \leqslant 74)$.

74.5 is the value you need to work with from writing either $< 75$ or $\leqslant 74$.

**c** $P(X \leqslant 40) \approx P\left(Z \leqslant \frac{40.5-60}{\sqrt{60}}\right)$
$= \Phi(-2.517) = 1 - 0.9941 = 0.0059$

**EXERCISE 2D**

1 $X \sim \text{Po}(42)$. Use the normal approximation to find:

**a** $P(X < 50)$ **b** $P(X \leqslant 40)$ **c** $P(X = 45)$

2 $X \sim \text{Po}(50)$. Use the normal approximation to find:

**a** $P(52 < X < 56)$ **b** $P(50 \leqslant X \leqslant 52)$

3 For a particular pond, in 1 ml of pond water, on average there are 117 microorganisms. Find the probability there are more than 1200 microorganisms in a 10 ml sample of the pond water.

M 4 On a particular train line, delayed trains occur at an average rate of eight per day.

a What is the probability that fewer than 100 trains are delayed on this line in a 14-day period?

b Decide if your Poisson model for the 14-day period is reasonable to use in this situation. Explain your decision.

PS 5 Fabio receives, on average, 48 emails at work each day. Emails are received at random and independently. Fabio can answer, at most, 60 emails each day. Find the probability that on a randomly chosen day, Fabio can answer all emails received.

6 Given that $X \sim \text{Po}(42)$ and $P(X > x) \leqslant 0.1$, find the minimum integer value of $x$.

PS 7 The number of tea lights that are lit at a place of memorial during one day follows a Poisson distribution with mean 38. How many tea lights should be available to be at least 98% certain that there are sufficient tea lights for the demand?

PS 8 The number of errors made by customers when using online banking transactions each week follows a Poisson distribution with mean 25.

a Find the probability that there are more than 32 customer errors in a randomly chosen week.

b Katya calculates that it is almost certain that the number of customer errors on a randomly chosen day is greater than 11 and less than 39. Use calculations to show how Katya arrived at her conclusion.



## 2.5 Hypothesis testing with the Poisson distribution

For a single observation from a population that has a Poisson distribution, we can directly compare Poisson probabilities or use a normal approximation to the Poisson distribution.

**REWIND**

Chapter 1 showed how to carry out a hypothesis test with the binomial distribution. To carry out hypothesis testing with the Poisson distribution, the process is the same. Go back and review Chapter 1 to remind yourself of the procedure involved.

### WORKED EXAMPLE 2.7

Records show that 1% of the population has a positive reaction to a test for a particular allergy. In a village, 120 people are tested and four people have a positive reaction. Test at the 5% level of significance if there is any evidence of an increase in the population with a positive reaction for this particular allergy.

**Answer**

$X \sim \text{B}(120, 0.01) \approx \text{Po}(1.2)$ — State the distribution; $n$ is large and $np < 5$ so approximate to Poisson.

$\text{H}_0: \lambda = 1.2$; $\text{H}_1: \lambda > 1.2$ — State the null and alternative hypotheses.

5% significance level one-tailed test — State the significance level of the test and whether one- or two-tailed.

$\text{P}(X \geqslant 4) = 1 - \text{P}(X \leqslant 3)$

$$= 1 - \text{e}^{-1.2}\left(\frac{1.2^0}{0!} + \frac{1.2^1}{1!} + \frac{1.2^2}{2!} + \frac{1.2^3}{3!}\right)$$

$$= 1 - 0.9662 = 3.38\%$$

Calculate the probability.

As $3.38\% < 5\%$, reject the null hypothesis. — Compare the probability with the significance level.

There is evidence at the 5% level of significance to suggest that in this village there has been an increase in the population who react positively to a test for this particular allergy. — Interpret the result.

**REWIND**

Using the binomial distribution, we can see the same conclusions would be reached:

$X \sim B(120, 0.01)$

$H_0 : \mu = 0.01$; $H_1 : \mu > 0.01$

5% level one-tailed test

$P(X \geqslant 4) = 1 - P(X \leqslant 3)$

$$= 1 - \left( \binom{120}{0} 0.01^0 \quad 0.99^{120} + \binom{120}{1} 0.01^1 \quad 0.99^{119} + \binom{120}{2} 0.01^2 \quad 0.99^{118} + \binom{120}{3} 0.01^3 \quad 0.99^{117} \right)$$

$= 1 - 0.9670 = 3.3\%$

As $3.3\% < 5\%$, reject the null hypothesis, as before.

**WORKED EXAMPLE 2.8**

Accidents on a stretch of road occur at the rate of seven each month. New traffic measures are put in place to reduce the number of accidents. In the following month, there are only two accidents.

**a** Test at the 5% level of significance if there is evidence that the new traffic measures have significantly reduced the number of accidents.

**b** Over a period of 6 months, there are 32 accidents. It is claimed the new traffic measures are no longer reducing the number of accidents.



**i** Test this claim at the 5% level of significance.

**ii** What would your conclusion be if you tested the claim at the 10% level of significance?

**Answer**

**a** $X \sim Po(7)$ — State the distribution.

$H_0$: $\lambda = 7$; $H_1$: $\lambda < 7$ — State the null and alternative hypotheses.

5% level one-tailed test — State the significance level of the test and whether one- or two-tailed.

$P(X \leqslant 2) = e^{-7}\left(\frac{7^0}{0!} + \frac{7^1}{1!} + \frac{7^2}{2!}\right)$

$= 0.0296 = 2.96\%$ — Calculate the probability.

As $2.96\% < 5\%$, reject the null hypothesis. — Compare the probability with the significance level.

There is evidence to suggest the new traffic measures reduce the number of accidents. — Interpret the result.

**b i** $X \sim Po(42) \approx N(42, 42)$ — State the distribution and the approximating distribution.

$H_0$: $\mu = 42$; $H_1$: $\mu < 42$ — State the null and alternative hypotheses; since we are approximating to the normal distribution, use $\mu$ and not $\lambda$.

5% level one-tailed test — State the significance level of the test and whether one- or two-tailed.

$P(X \leqslant 32) = P\left(Z \leqslant \frac{32.5 - 42}{\sqrt{42}}\right)$

$= \Phi(-1.466) = 1 - 0.9287 = 0.0713$ — Calculate the probability. Remember to use the continuity correction.

As $7.13\% > 5\%$, accept the null hypothesis. — Compare the probability with the significance level.

There is insufficient evidence to suggest the new traffic measures have reduced the number of accidents. — Interpret the result.

ii $7.31\% < 10\%$

There is evidence at the 10% significance level that the new traffic measures reduce the number of accidents.

Show the relevant comparison; there is no need to carry out an additional calculation.

Interpret the result.

## EXPLORE 2.5

For Worked example 2.8, at a 5% level of significance results from the first month suggest there is evidence that the new traffic measures reduce the number of accidents, yet results from 6 months suggest there is insufficient evidence. Which is true? What other factors might affect the result? At a 5% significance level, statistically 1 month in every 20 months there will be fewer accidents. What does it imply if the number of accidents in the first month is a statistical fluke? If drivers become more careless over time as they get used to a new road layout, what does this imply about your answer to part **b ii**? Over what period of time should you collect data to test the effect of the new traffic measures? What else can you do to show if the new traffic measures have reduced the number of accidents?



## EXERCISE 2E

 PS **1** The number of calls to a consumer hotline can be modelled by a Poisson distribution with mean 62 calls every 5 minutes. Salina believes this average is too low and observes the number of calls recorded during a randomly chosen 5-minute interval to be 70. Stating the null and alternative hypotheses, test Salina's belief at the 10% significance level.

PS **2** A small shop sells, on average, seven laptops per week. Following a price rise, the number of laptops sold drops to four per week. Test at the 5% significance level whether the sales of laptops have significantly reduced.

PS **3** At a certain company, machine faults occur randomly and at a constant mean rate of 1.5 per week. Following an overhaul of the machines, the company boss wishes to determine if the mean rate of machine faults has fallen. The number of machine faults recorded over 26 weeks is 28. Use a suitable approximation and test at the 5% significance level whether the mean rate has fallen.

PS **4** The number of misprints per page of a newspaper follows a Poisson distribution with mean two per page. Following new procedures, 49 misprints are found in 32 pages of the newspaper.

**a** Use a suitable approximation to test at the 5% level of significance if the mean number of misprints has changed.

**b** How many misprints would be needed for a Type I error to have occurred?

 **REWIND**

Hypothesis testing based on direct evaluation of Poisson probabilities of making Type I and Type II errors are calculated in the same way as for the binomial distribution in Chapter 1.

## Checklist of learning and understanding

- A Poisson distribution is a suitable model to use for events that occur:
  - singly
  - independently
  - at random in a given interval of time or space
  - at a constant rate: this is the mean number of events in a given interval that is proportional to the size of the interval.
- If for a Poisson distribution $X \sim \text{Po}(\lambda)$, where $\lambda > 0$, then:
  - $P(X = r) = e^{-\lambda} \times \frac{\lambda^r}{r!}$, where $r = 0, 1, 2, 3, \ldots$
  - Mean $E(X) = \lambda$ and Variance $\text{Var}(X) = \lambda$.
- A binomial distribution $B(n, p)$, where $n$ is large such that $n > 50$, and $p$ is small such that $np < 5$, can be approximated by a Poisson distribution $\text{Po}(np)$. The larger the value for $n$ and the smaller the value for $p$, the better the approximation.
- A Poisson distribution $\text{Po}(\lambda)$, where $\lambda > 15$, may be approximated by the normal distribution $N(\lambda, \lambda)$. A continuity correction must be applied.

## END-OF-CHAPTER REVIEW EXERCISE 2

1 A shop sells a particular item at the rate of four per week. Assuming that items are sold independently and at random, find the probability of the following events.

a Exactly three items are sold in a randomly chosen week. [2]

b Three or more items are sold in a randomly chosen week. [3]

c Exactly six items are sold in a randomly chosen 2-week period. [2]

d Six or more items are sold in a randomly chosen 2-week period. [3]

M 2 The numbers of email enquiries received by an online retailer over a period of 160 days are as follows.

| **Emails per day** | 0 | 1 | 2 | 3 | 4 | 5 | 6 | 7 | 8 | $\geqslant 9$ |
|---|---|---|---|---|---|---|---|---|---|---|
| **Frequency** | 7 | 18 | 24 | 36 | 32 | 22 | 13 | 6 | 2 | 0 |

a Find the mean and variance and use these to comment on whether the data are consistent with a Poisson distribution. What other values could you calculate to help you decide? [4]

b Assuming that the number of emails follows a Poisson distribution with mean 3.4, what is the probability of receiving between four and eight emails, inclusive, on any particular day? [3]

PS 3 Threats to steal online data are detected by an IT security firm independently and at random at the rate of five per day. Use a suitable approximating distribution where appropriate and find the probability of the following events.

a Exactly 25 threats are detected in a 5-day period. [3]

b More than 60 threats are detected over a 10-day period. [4]

c 140 to 152 (inclusive) threats are detected over a 30-day period. [4]

PS 4 Pairs of denim jeans made at a factory are found to have faults at an average rate of one in every 100. The faults occur independently and at random. In a batch of 300 pairs of denim jeans produced by a new employee, five are found to have faults. Show that at the 5% level of significance the pairs of denim jeans produced by the new employee do not have more faults than expected. [6]

PS 5 An airline has found, from long experience, that the number of people booked on flights who do not arrive at the airport to catch the flight follows a Poisson distribution at an average rate of 2% per flight. For a flight with 146 available seats, 150 seats are sold. Use a suitable approximating distribution to find the probability that there are sufficient seats available for everyone who arrives to catch the flight. [4]

M 6 At a certain shop the demand for hair dryers has a Poisson distribution with mean 3.4 per week.

i Find the probability that, in a randomly chosen two-week period, the demand is for exactly 5 hair dryers. [3]

ii At the beginning of a week the shop has a certain number of hair dryers for sale. Find the probability that the shop has enough hair dryers to satisfy the demand for the week if:

a they have 4 hair dryers in the shop [2]

b they have 5 hair dryers in the shop. [2]

iii Find the smallest number of hair dryers that the shop needs to have at the beginning of a week so that the probability of being able to satisfy the demand that week is at least 0.9. [3]

*Cambridge International AS & A Level Mathematics 9709 Paper 72 Q6 June 2016*

7 A Lost Property office is open 7 days a week. It may be assumed that items are handed in to the office randomly, singly and independently.

i State another condition for the number of items handed in to have a Poisson distribution. **[1]**

It is now given that the number of items handed in per week has the distribution Po(4.0).

ii Find the probability that exactly 2 items are handed in on a particular day. **[2]**

iii Find the probability that at least 4 items are handed in during a 10-day period. **[3]**

iv Find the probability that, during a certain week, 5 items are handed in altogether, but no items are handed in on the first day of the week. **[3]**

*Cambridge International AS & A Level Mathematics 9709 Paper 73 Q7 June 2014*

8 The number of radioactive particles emitted per 150-minute period by some material has a Poisson distribution with mean 0.7.

i Find the probability that at most 2 particles will be emitted during a randomly chosen 10-hour period. **[3]**

ii Find, in minutes, the longest time period for which the probability that no particles are emitted is at least 0.99. **[5]**

*Cambridge International AS & A Level Mathematics 9709 Paper 71 Q4 November 2013*

9 Customers arrive at an enquiry desk at a constant average rate of 1 every 5 minutes.



i State one condition for the number of customers arriving in a given period to be modelled by a Poisson distribution. **[1]**

Assume now that a Poisson distribution is a suitable model.

ii Find the probability that exactly 5 customers will arrive during a randomly chosen 30-minute period. **[2]**

iii Find the probability that fewer than 3 customers will arrive during a randomly chosen 12-minute period. **[3]**

iv Find an estimate of the probability that fewer than 30 customers will arrive during a randomly chosen 2-hour period. **[4]**

*Cambridge International AS & A Level Mathematics 9709 Paper 71 Q6 November 2011*

10 The number of goals scored per match by Everly Rovers is represented by the random variable $X$, which has mean 1.8.

i State two conditions for $X$ to be modelled by a Poisson distribution. **[2]**

Assume now that $X \sim \mathrm{Po}(1.8)$.

ii Find $\mathrm{P}(2 < X < 6)$. **[2]**

iii The manager promises the team a bonus if they score at least 1 goal in each of the next 10 matches. Find the probability that they win the bonus. **[3]**

*Cambridge International AS & A Level Mathematics 9709 Paper 72 Q3 June 2011*

**11** In the past, the number of house sales completed per week by a building company has been modelled by a random variable which has the distribution Po(0.8). Following a publicity campaign, the builders hope that the mean number of sales per week will increase. In order to test at the 5% significance level whether this is the case, the total number of sales during the first 3 weeks after the campaign is noted. It is assumed that a Poisson model is still appropriate.

i Given that the total number of sales during the 3 weeks is 5, carry out the test. **[6]**

ii During the following 3 weeks the same test is carried out again, using the same significance level. Find the probability of a Type I error. **[3]**

iii Explain what is meant by a Type I error in this context. **[1]**

iv State what further information would be required in order to find the probability of a Type II error. **[1]**

*Cambridge International AS & A Level Mathematics 9709 Paper 73 Q7 November 2010*

**12** A hospital patient's white blood cell count has a Poisson distribution. Before undergoing treatment the patient had a mean white blood cell count of 5.2. After the treatment a random measurement of the patient's white blood cell count is made, and is used to test at the 10% significance level whether the mean white blood cell count has decreased.

i State what is meant by a Type I error in the context of the question, and find the probability that the test results in a Type I error. **[4]**

ii Given that the measured value of the white blood cell count after the treatment is 2, carry out the test. **[3]**

iii Find the probability of a Type II error if the mean white blood cell count after the treatment is actually 4.1. **[3]**

*Cambridge International AS & A Level Mathematics 9709 Paper 71 Q7 June 2010*

**13** Major avalanches can be regarded as randomly occurring events. They occur at a uniform average rate of 8 per year.

i Find the probability that more than 3 major avalanches occur in a 3-month period. **[3]**

ii Find the probability that any two separate 4-month periods have a total of 7 major avalanches. **[3]**

iii Find the probability that a total of fewer than 137 major avalanches occur in a 20-year period. **[4]**

*Cambridge International AS & A Level Mathematics 9709 Paper 71 Q3 June 2009*

# Chapter 3
# Linear combinations of random variables

**In this chapter you will learn how to:**

- find means and variances of linear combinations of random variables
- calculate probabilities of linear combinations of random variables
- solve problems involving linear combinations of random variables.

**PREREQUISITE KNOWLEDGE**

| Where it comes from | What you should be able to do | Check your skills |
|---|---|---|
| Probability & Statistics 1, Chapters 6 and 7 | Find the mean and variance of a discrete probability distribution. | **1** A fair six-sided die is numbered 1, 1, 2, 3, 3, 3. Let $X$ be the score when the die is rolled once. Find the mean $E(X)$ and variance $Var(X)$. |
| Probability & Statistics 1, Chapter 8 | Use normal distribution tables to calculate probabilities. | **2** Given that $X \sim N(22, 5)$, find: **a** $P(X < 23)$ **b** $P(X > 21)$ |
| Chapter 2 | Find probabilities using the Poisson distribution. | **3** Given that $X \sim Po(5.4)$, find: **a** $P(X = 3)$ **b** $P(X \leqslant 1)$ |

## Why do we study linear combinations of random variables?

The Probability & Statistics 1 Coursebook described how to find the expectation (mean) and variance for a discrete random variable, and the binomial distribution. Chapter 2 of this book explained the Poisson distribution. The Probability & Statistics 1 Coursebook also explored solving problems using a continuous random variable, the normal distribution. In each example involving the binomial, Poisson or normal distributions, the information given enabled the calculation of probabilities, and mean and variance for that random variable. However, in some problem-solving situations it is more appropriate to use a combination of random variables. For example, in the monthly profit or loss of a stock portfolio, the investments in stocks and shares in several companies are given as a single value. This value is actually the sum of profits and losses of the individual investments. A triathlon time is a combination of the times taken to complete the different events, and the results of each event are independent random variables. The weight of a jar of honey is a combination of the weight of the honey, the weight of the jar and the weight of the lid, which is a combination of independent random variables.

In this chapter, you will solve problems using the expectation and variance of linear combinations of random variables.

## 3.1 Expectation and variance

Xing has a fair six-sided die, numbered 1, 1, 2, 2, 2, 4. Let the random variable $X$ be 'the score obtained when the die is rolled'; then the probability distribution for $X$ is shown in the following table.

| $x$ | 1 | 2 | 4 |
|---|---|---|---|
| $P(X = x)$ | $\frac{1}{3}$ | $\frac{1}{2}$ | $\frac{1}{6}$ |

$$E(X) = \left(1 \times \frac{1}{3}\right) + \left(2 \times \frac{1}{2}\right) + \left(4 \times \frac{1}{6}\right) = 2$$

$$Var(X) = \left(1^2 \times \frac{1}{3}\right) + \left(2^2 \times \frac{1}{2}\right) + \left(4^2 \times \frac{1}{6}\right) - 2^2 = 1$$

Yaffa has a fair six-sided die, numbered 4, 4, 5, 5, 5, 7. Let the random variable $Y$ be 'the score obtained when the die is rolled'; then the probability distribution for $Y$ is shown in the following table.

| $y$ | 4 | 5 | 7 |
|---|---|---|---|
| $P(Y = y)$ | $\frac{1}{3}$ | $\frac{1}{2}$ | $\frac{1}{6}$ |

$$E(Y) = \left(4 \times \frac{1}{3}\right) + \left(5 \times \frac{1}{2}\right) + \left(7 \times \frac{1}{6}\right) = 5$$

$$Var(X) = \left(4^2 \times \frac{1}{3}\right) + \left(5^2 \times \frac{1}{2}\right) + \left(7^2 \times \frac{1}{6}\right) - 5^2 = 1$$

Compare the expectation and variance for $X$ and $Y$. What do you notice?

The numbers on Yaffa's die are all 3 more than the numbers on Xing's die. The mean score for Yaffa's die is 3 more than the mean score on Xing's die, but the variance for the two dice is the same. Why should this happen?

Consider the relative position of the dice scores on a number line.

1 1 2 2 2 4 Xing's die

4 4 5 5 5 7 Yaffa's die

We can see that the numbers on Yaffa's die are all +3 compared to the numbers on Xing's die, meaning the mean score will also be +3. The variability between the set of numbers on each die – that is, their position on the number line relative to each other – remains the same; hence, the variances are the same.

**EXPLORE 3.1**



Would adding the same value to each number of Mo's fair six-sided die, numbered 0, 0, 1, 1, 1, 3, have the same effect on the mean and variance? Find the mean and variance for Mo's die, then add or subtract the same number to each of the numbers on the die and find the mean and variance for the new set of numbers. Do you get the same effect as for Yaffa's die? Explore these results further with an example of your own choosing.

**KEY POINT 3.1**

For a random variable $X$ and constant $b$:

$E(X + b) = E(X) + b$ and $Var(X + b) = Var(X)$

**REWIND**

Remember that mean$(x - b)$ = mean$(x) - b$ and var$(x - b)$ = var$(x)$, which were covered in Chapter 2 and Chapter 3, respectively, of the Probability & Statistics Coursebook 1.

Quenby has a fair six-sided die, numbered 2, 2, 4, 4, 4, 8.

Let the random variable $Q$ be 'the score obtained when the die is rolled'; then the probability distribution for $Q$ is:

| $n$ | 2 | 4 | 8 |
|---|---|---|---|
| $P(Q = n)$ | $\frac{1}{3}$ | $\frac{1}{2}$ | $\frac{1}{6}$ |

$$E(Q) = \left(2 \times \frac{1}{3}\right) + \left(4 \times \frac{1}{2}\right) + \left(8 \times \frac{1}{6}\right) = 4$$

$$Var(Q) = \left(2^2 \times \frac{1}{3}\right) + \left(4^2 \times \frac{1}{2}\right) + \left(8^2 \times \frac{1}{6}\right) - 4^2 = 4$$

Compare the expectation and variance for $X$ and $Q$. What do you notice?

The numbers on Quenby's die are double the numbers on Xing's die, and the expectation for Quenby's die is also double the expectation for Xing's die, whereas the variance has quadrupled.

| | | | | |
|---|---|---|---|---|
| 1 1 | 2 2 2 | 4 | | Xing's die |
| | 2 2 | 4 4 4 | 8 | Quenby's die |

Looking at the position of the numbers on a number line, we can see that if the numbers on a die are doubled, the positions of the numbers relative to each other also increases. In fact $\text{Var}(Q) = 4 \times \text{Var}(X) = 2^2 \times \text{Var}(X)$.

**KEY POINT 3.2**

For a random variable $X$ and constant $a$:

$\text{E}(aX) = a\text{E}(X)$ and $\text{Var}(aX) = a^2\text{Var}(X)$

**EXPLORE 3.2**

If we put the results in Key point 3.1 and Key point 3.2 together, we have:

$$\text{E}(X + b) = \text{E}(X) + b \text{ and } \text{E}(aX) = a\text{E}(X)$$

$$\text{Var}(X + b) = \text{Var}(X) \text{ and } \text{Var}(aX) = a^2\text{Var}(X)$$

Then we have the following results:

$$\text{E}(aX + b) = a\text{E}(X) + b \text{ and } \text{Var}(aX + b) = a^2\text{Var}(X)$$

Choose a rule to create your own set of numbers for a die to check the generalisations above. For example, starting with the numbers on Xing's die, multiplying by 3 and adding 1 we have $3X + 1$ and the numbers on the die are 4, 4, 7, 7, 7, 13. What is the mean and variance? Will we get the same values for the mean and variance if we use the rules given above?

Try a different rule for altering the numbers on Yaffa's die. What do you find?

**KEY POINT 3.3**

For a random variable $X$ and constants $a$ and $b$:

$\text{E}(aX + b) = a\text{E}(X) + b$ and $\text{Var}(aX + b) = a^2\text{Var}(X)$

**WORKED EXAMPLE 3.1**

Given that a random variable $X$ has the probability distribution shown in the following table, find:

**a** $E(2X+3)$ **b** $Var(2X+3)$

| $x$ | 2 | 3 | 5 | 7 |
|---|---|---|---|---|
| $P(X = x)$ | 0.4 | 0.1 | 0.3 | 0.2 |

**Answer**

$E(X) = (2 \times 0.4) + (3 \times 0.1) + (5 \times 0.3) + (7 \times 0.2) = 4$

$Var(X) = (2^2 \times 0.4) + (3^2 \times 0.1) + (5^2 \times 0.3) + (7^2 \times 0.2) - 4^2$
$= 19.8 - 16 = 3.8$

First, find $E(X)$ and $Var(X)$.

**a** $E(2X+3) = 2E(X) + 3 = 2 \times 4 + 3 = 11$

Use the expectation $E(X) = 4$ with the general result $E(aX+b) = aE(X) + b$.

**b** $Var(2X+3) = 2^2 Var(X) = 4 \times 3.8 = 15.2$

Use the variance $Var(X) = 3.8$ with the general result $Var(aX+b) = a^2 Var(X)$.

**WORKED EXAMPLE 3.2**



The random variable $X$ is the number of tails obtained when three fair coins are tossed.

**a** Construct the probability distribution table for $X$ and find $E(X)$ and $Var(X)$.

**b** Find:

**i** $E(2X+1)$ **ii** $Var(2X+1)$

**Answer**

**a**

| $x$ | 0 | 1 | 2 | 3 |
|---|---|---|---|---|
| $P(X = x)$ | $\frac{1}{8}$ | $\frac{3}{8}$ | $\frac{3}{8}$ | $\frac{1}{8}$ |

$E(X) = \frac{3}{8} + \frac{6}{8} + \frac{3}{8} = 1\frac{1}{2}$

$Var(X) = \frac{3}{8} + \frac{12}{8} + \frac{9}{8} - \left(1\frac{1}{2}\right)^2 = \frac{3}{4}$

**b i** $E(2X+1) = 2E(X) + 1 = \left(2 \times 1\frac{1}{2}\right) + 1 = 4$

Use the expectation $E(X) = 1\frac{1}{2}$.

**ii** $Var(2X+1) = 2^2 Var(X) = 4 \times \frac{3}{4} = 3$

Use the variance $Var(X) = \frac{3}{4}$.

The random variable in Worked example 3.2 can also be expressed using notation for the binomial distribution $X \sim B\left(3, \frac{1}{2}\right)$. Using the properties of a binomial distribution we have $E(X) = np = \frac{3}{2}$ and $Var(X) = npq = \frac{3}{4}$. Notice these are the same values as those obtained from the table of values in part **a** of the Worked example. Therefore, we can conclude that Key point 3.3 also applies to random variables that have a binomial distribution.

### WORKED EXAMPLE 3.3

Given that the random variable $X \sim B(8, 0.3)$, find:

**a** $E(4X - 1)$

**b** $Var(4X - 1)$

**Answer**

$E(X) = np = 8 \times 0.3 = 2.4$ — First, find $E(X)$ and $Var(X)$.

$Var(X) = npq = 8 \times 0.3 \times (1 - 0.3) = 1.68$

**a** $E(4X - 1) = 4E(X) - 1 = (4 \times 2.4) - 1 = 8.6$ — Use the expectation $E(X) = 2.4$.

**b** $Var(4X - 1) = 4^2 Var(X) = 16 \times 1.68 = 26.88$ — Use the variance $Var(X) = 1.68$.

### EXERCISE 3A

**1** The random variable $X \sim B(25, 0.6)$. Find:

**a** $E(5X - 1)$ **b** $Var(5X - 1)$

**2** The random variable $X \sim B(30, 0.2)$. Find:

**a** $E(2X + 3)$ **b** $Var(2X + 3)$

M **3** The random variable $X$ has expectation 2.4 and variance 0.8.

**a** Find:

**i** $E(3X + 2)$ **ii** $Var(3X + 2)$

**b** Find two pairs of values for the constants $a$ and $b$ such that $E(aX + b) = 32$ and $Var(aX + b) = 20$.

M **4** The random variable $X$ has expectation 5 and variance 1.3.

**a** Find:

**i** $E(4X - 1)$ **ii** $Var(4X - 1)$

**b** Find two pairs of values for the constants $a$ and $b$ such that $E(aX - b) = 20$ and $Var(aX - b) = 130$.

**5** The random variable $X$ can take the values 10, 20 and 30, with corresponding probabilities 0.5, 0.3 and 0.2. Find:

**a** $E\left(\frac{1}{2}X\right)$ **b** $Var\left(\frac{1}{2}X\right)$

**6** $X$ is the score from a single roll of an ordinary fair six-sided die. Find:

**a** $E(4X - 3)$ **b** $Var(4X - 3)$

**7** The random variable $X$ has the probability distribution shown in the following table.

| $x$ | 6 | 28 |
|---|---|---|
| $P(X = x)$ | 0.8 | 0.2 |

Find:

**a** $E(2X + 5)$ **b** $Var(2X + 5)$

**8** The random variable $X$ has mean $\mu$ and variance $\sigma^2$. Find two pairs of values for the constants $a$ and $b$ such that $E(aX - b) = 0$ and $Var(aX - b) = 1$.

**9** Over a long period, the average temperature in an area of Mauritius is 24.5 °C. The variance in temperature is 2 °C. Using the conversion F = 1.8C + 32, where F is the temperature in degrees Fahrenheit (°F) and C is the temperature in degrees Celsius (°C), work out the average temperature and standard deviation in temperature in degrees Fahrenheit.

## 3.2 Sum and difference of independent random variables

Wendy has two tetrahedral dice: a green die numbered 1, 1, 2, 3 and a blue die numbered 1, 1, 2, 2.

What will be the mean and variance of the total when the scores on the two dice are added together? How are these related to the mean and variance of the scores on the two individual dice?



Let the random variable $G$ be the score on the green die. Then $E(G) = 1\frac{3}{4}$ and $Var(G) = \frac{11}{16}$.

Let the random variable $B$ be the score on the blue die. Then $E(B) = 1\frac{1}{2}$ and $Var(B) = \frac{1}{4}$.

The table shows the possible outcomes of rolling both dice and adding the scores.

| + | 1 | 1 | 2 | 3 |
|---|---|---|---|---|
| 1 | 2 | 2 | 3 | 4 |
| 1 | 2 | 2 | 3 | 4 |
| 2 | 3 | 3 | 4 | 5 |
| 2 | 3 | 3 | 4 | 5 |

Let the random variable $W$ be the sum of scores on Wendy's green and blue dice.

| $w$ | 2 | 3 | 4 | 5 |
|---|---|---|---|---|
| $P(W = w)$ | $\frac{1}{4}$ | $\frac{3}{8}$ | $\frac{1}{4}$ | $\frac{1}{8}$ |

$$E(W) = \left(2\times\frac{1}{4}\right)+\left(3\times\frac{3}{8}\right)+\left(4\times\frac{1}{4}\right)+\left(5\times\frac{1}{8}\right) = 3\frac{1}{4}$$

$$Var(W) = \left(2^2\times\frac{1}{4}\right)+\left(3^2\times\frac{3}{8}\right)+\left(4^2\times\frac{1}{4}\right)+\left(5^2\times\frac{1}{8}\right) - 3\frac{1}{4}^2 = \frac{15}{16}$$

Comparing these values with the expectation and variance for the random variables $G$ and $B$ we find that:

$$E(G) + E(B) = 1\tfrac{3}{4} + 1\tfrac{1}{2} = 3\tfrac{1}{4} \text{ and } Var(G) + Var(B) = \frac{11}{16} + \frac{1}{4} = \frac{15}{16}.$$

**KEY POINT 3.4**

For two independent random variables $X$ and $Y$:

$\mathrm{E}(X+Y) = \mathrm{E}(X) + \mathrm{E}(Y)$ and $\mathrm{Var}(X+Y) = \mathrm{Var}(X) + \mathrm{Var}(Y)$

**EXPLORE 3.3**

- For the independent random variables $X$, $Y$ and $N$, would the results in Key point 3.4 be true for the difference between two independent random variables?
- Using the random variables $X$ and $Y$ from Section 3.1, or by making up your own, draw a probability table for $X - Y$ and work out $\mathrm{E}(X-Y)$ and $\mathrm{Var}(X-Y)$. What do you notice?
- For the independent random variables $X$, $Y$ and $N$, would the results in Key point 3.4 be true for the sum of more than two independent random variables?
- Draw a probability table for $X+Y+N$ and work out $\mathrm{E}(X+Y+N)$ and $\mathrm{Var}(X+Y+N)$. What do you notice?

What about multiples of independent random variables?

Suppose the numbers on the green die are doubled and the numbers on the blue die are tripled. The table shows the possible outcomes of rolling both dice and adding the scores.



| + | 2 | 2 | 4 | 6 |
|---|---|---|---|---|
| 3 | 5 | 5 | 7 | 9 |
| 3 | 5 | 5 | 7 | 9 |
| 6 | 8 | 8 | 10 | 12 |
| 6 | 8 | 8 | 10 | 12 |

Let the random variable $D$ be such that $D = 2G + 3B$.

| $d$ | 5 | 7 | 8 | 9 | 10 | 12 |
|---|---|---|---|---|---|---|
| $\mathrm{P}(D=d)$ | $\frac{1}{4}$ | $\frac{1}{8}$ | $\frac{1}{4}$ | $\frac{1}{8}$ | $\frac{1}{8}$ | $\frac{1}{8}$ |

From the table we can work out $\mathrm{E}(D)$ and $\mathrm{Var}(D)$.

$$\mathrm{E}(D) = \left(5\times\frac{1}{4}\right)+\left(7\times\frac{1}{8}\right)+\left(8\times\frac{1}{4}\right)+\left(9\times\frac{1}{8}\right)+\left(10\times\frac{1}{8}\right)+\left(12\times\frac{1}{8}\right) = 8$$

$$\mathrm{Var}(D) = \left(5^2\times\frac{1}{4}\right)+\left(7^2\times\frac{1}{8}\right)+\left(8^2\times\frac{1}{4}\right)+\left(9^2\times\frac{1}{8}\right)+\left(10^2\times\frac{1}{8}\right)+\left(12^2\times\frac{1}{8}\right) - 8^2 = 5$$

And if we use the expectation and variance for the random variables $G$ and $B$, we find that:

$$\mathrm{E}(2G+3B) = 2\mathrm{E}(G) + 3\mathrm{E}(B) = 2\times1\tfrac{3}{4} + 3\times1\tfrac{1}{2} = 3\tfrac{1}{2} + 4\tfrac{1}{2} = 8 = \mathrm{E}(D)$$

$$\mathrm{Var}(2G+3B) = 2^2\mathrm{Var}(G) + 3^2\mathrm{Var}(B) = 2^2\times\frac{11}{16} + 3^2\times\frac{1}{4} = 2\tfrac{3}{4} + 2\tfrac{1}{4} = 5 = \mathrm{Var}(D)$$

**KEY POINT 3.5**

For two independent random variables $X$ and $Y$ and constants $a$ and $b$:

$\mathrm{E}(aX + bY) = a\mathrm{E}(X) + b\mathrm{E}(Y)$ and $\mathrm{Var}(aX + bY) = a^2\mathrm{Var}(X) + b^2\mathrm{Var}(Y)$

These results can be extended to any number of independent random variables.

**WORKED EXAMPLE 3.4**

The following tables give the probability distributions for two independent random variables, $X$ and $Y$.

| $x$ | 2 | 3 | 5 |
|---|---|---|---|
| $\mathrm{P}(X = x)$ | 0.2 | 0.3 | 0.5 |

| $y$ | 0 | 1 |
|---|---|---|
| $\mathrm{P}(Y = y)$ | 0.6 | 0.4 |

**a** Draw a probability distribution table for the random variable $S$, where $S = X + Y$, and find $\mathrm{E}(S)$ and $\mathrm{Var}(S)$.

**b** Draw a probability distribution table for the random variable $T$, where $T = X - Y$, and find $\mathrm{E}(T)$ and $\mathrm{Var}(T)$.

**c** $\mathrm{E}(X) = 3.8$, $\mathrm{Var}(X) = 1.56$; $\mathrm{E}(Y) = 0.4$, $\mathrm{Var}(Y) = 0.24$. Comment on your results for expectation and variance of the sum and difference of two independent random variables.

**Answer**

**a**

| + | 2 | 3 | 5 |
|---|---|---|---|
| 0 | 2 | 3 | 5 |
| 1 | 3 | 4 | 6 |

| | 2 | 3 | 5 |
|---|---|---|---|
| 0 | $0.6 \times 0.2$ | $0.6 \times 0.3$ | $0.6 \times 0.5$ |
| 1 | $0.4 \times 0.2$ | $0.4 \times 0.3$ | $0.4 \times 0.5$ |

Draw a table showing possible outcomes for $X + Y$; then a second table showing probabilities of these outcomes.

| $s$ | 2 | 3 | 4 | 5 | 6 |
|---|---|---|---|---|---|
| $\mathrm{P}(S = s)$ | 0.12 | 0.26 | 0.12 | 0.3 | 0.2 |

The probability distribution table for $S$ summarises the information from both tables.

$\mathrm{E}(S) = (2 \times 0.12) + (3 \times 0.26) + (4 \times 0.12) + (5 \times 0.3) + (6 \times 0.2) = 4.2$

$\mathrm{Var}(S) = (2^2 \times 0.12) + (3^2 \times 0.26) + (4^2 \times 0.12) + (5^2 \times 0.3) + (6^2 \times 0.2) - 4.2^2$
$= 19.44 - 17.64 = 1.8$

Use your knowledge from Probability & Statistics Coursebook 1 to find $\mathrm{E}(S)$ and $\mathrm{Var}(S)$.

**b**

| − | 2 | 3 | 5 |
|---|---|---|---|
| 0 | 2 | 3 | 5 |
| 1 | 1 | 2 | 4 |

| | 2 | 3 | 5 |
|---|---|---|---|
| 0 | $0.6 \times 0.2$ | $0.6 \times 0.3$ | $0.6 \times 0.5$ |
| 1 | $0.4 \times 0.2$ | $0.4 \times 0.3$ | $0.4 \times 0.5$ |

Draw a table showing possible outcomes for $X - Y$; then a second table showing probabilities of these outcomes.

| $t$ | 1 | 2 | 3 | 4 | 5 |
|---|---|---|---|---|---|
| $\mathrm{P}(T = t)$ | 0.08 | 0.24 | 0.18 | 0.2 | 0.3 |

The probability distribution table for $T$ summarises the information from both tables.

$$\mathrm{E}(T) = (1 \times 0.08) + (2 \times 0.24) + (3 \times 0.18) + (4 \times 0.2) + (5 \times 0.3) = 3.4$$

$$\mathrm{Var}(T) = (1^2 \times 0.08) + (2^2 \times 0.24) + (3^2 \times 0.18) + (4^2 \times 0.2) + (5^2 \times 0.3) - 3.4^2$$
$$= 13.36 - 11.56 = 1.8$$

**c** When we sum values of independent random variables, we add the expectations and add the variances:

$$\mathrm{E}(S) = \mathrm{E}(X + Y) = \mathrm{E}(X) + \mathrm{E}(Y),$$
$$\mathrm{Var}(S) = \mathrm{Var}(X + Y) = \mathrm{Var}(X) + \mathrm{Var}(Y).$$

This is the result you should have found in the first part of Explore 3.3.

$$\mathrm{E}(T) = \mathrm{E}(X - Y) = \mathrm{E}(X) - \mathrm{E}(Y)$$
$$\mathrm{Var}(T) = \mathrm{Var}(X - Y) = \mathrm{Var}(X) + \mathrm{Var}(Y)$$

**TIP**

When independent random variables are combined, their variances are always added.

Note that this example shows that, to find the difference of independent random variables, we find the difference in the expectation, but add the variances.

**WORKED EXAMPLE 3.5**

The random variable $X \sim \mathrm{B}(12, 0.2)$ and the random variable $Y \sim \mathrm{B}(2, 0.8)$.

The random variable $W = X - Y$. Work out the expectation and variance of $W$.

**Answer**

$$\mathrm{E}(X) = 12 \times 0.2 = 2.4$$
$$\mathrm{Var}(X) = 12 \times 0.2 \times (1 - 0.2) = 1.92$$
$$\mathrm{E}(Y) = 2 \times 0.8 = 1.6$$
$$\mathrm{Var}(Y) = 2 \times 0.8 \times 0.2 = 0.32$$

Use your knowledge from Probability & Statistics 1 Coursebook and Chapter 2 to work out the expectation and variance of $X$ and $Y$.

$$\mathrm{E}(W) = \mathrm{E}(X - Y) = \mathrm{E}(X) - \mathrm{E}(Y)$$
$$= 2.4 - 1.6 = 0.8$$
$$\mathrm{Var}(W) = \mathrm{Var}(X - Y)$$
$$= \mathrm{Var}(X) + \mathrm{Var}(Y)$$
$$= 1.92 + 0.32 = 2.24$$

The tip from the previous worked example: will help you to work out the expectation and variance of $W$.



## When is $2X \neq X + X$?

Let us look again at Xing's six-sided die, numbered 1, 1, 2, 2, 2, 4, from Section 3.1. We learned that $\mathrm{E}(X) = 2$ and $\mathrm{Var}(X) = 1$.

Quenby's die, also from Section 3.1, has numbers 2, 2, 4, 4, 4, 8. We know that $\mathrm{E}(Q) = 4$ and $\mathrm{Var}(Q) = 4$.

The numbers on Quenby's die are double the numbers on Xing's die, so $Q = 2X$.

$\mathrm{E}(Q) = \mathrm{E}(2X) = 2\mathrm{E}(X) = 2 \times 2 = 4$ and $\mathrm{Var}(Q) = \mathrm{Var}(2X) = 2^2\mathrm{Var}(X) = 2^2 \times 1 = 4$, as expected.

What would happen if we rolled Xing's die twice and added the scores? We do not get the outcomes of doubling the score; this is not the same as $2X$ from one roll of the die. In this situation we say that $X_1$ and $X_2$ are two independent outcomes from rolling Xing's die twice and we write $W = X_1 + X_2$.

To explore how this works, first draw a table of the outcomes:

| + | 1 | 1 | 2 | 2 | 2 | 4 |
|---|---|---|---|---|---|---|
| 1 | 2 | 2 | 3 | 3 | 3 | 5 |
| 1 | 2 | 2 | 3 | 3 | 3 | 5 |
| 2 | 3 | 3 | 4 | 4 | 4 | 6 |
| 2 | 3 | 3 | 4 | 4 | 4 | 6 |
| 2 | 3 | 3 | 4 | 4 | 4 | 6 |
| 4 | 5 | 5 | 6 | 6 | 6 | 8 |

The probability distribution table for $W$ is shown in the following table.

| $w$ | 2 | 3 | 4 | 5 | 6 | 8 |
|---|---|---|---|---|---|---|
| $P(W = w)$ | $\frac{1}{9}$ | $\frac{1}{3}$ | $\frac{1}{4}$ | $\frac{1}{9}$ | $\frac{1}{6}$ | $\frac{1}{36}$ |

$$E(W) = \left(2 \times \frac{1}{9}\right) + \left(3 \times \frac{1}{3}\right) + \left(4 \times \frac{1}{4}\right) + \left(5 \times \frac{1}{9}\right) + \left(6 \times \frac{1}{6}\right) + \left(8 \times \frac{1}{36}\right) = 4$$

$$\text{Var}(W) = \left(2^2 \times \frac{1}{9}\right) + \left(3^2 \times \frac{1}{3}\right) + \left(4^2 \times \frac{1}{4}\right) + \left(5^2 \times \frac{1}{9}\right) + \left(6^2 \times \frac{1}{6}\right) + \left(8^2 \times \frac{1}{36}\right) - 4^2 = 2$$

Using Key point 3.4, we find that $E(W) = E(X_1) + E(X_2) = 2 + 2 = 4$ and $\text{Var}(W) = \text{Var}(X_1) + \text{Var}(X_2) = 1 + 1 = 2$, as expected.

**TIP**

You need to be able to distinguish between a random variable $2X$, which refers to twice the size of an observation from the random variable $X$, and $X_1 + X_2$, which refers to the sum of two independent observations of the random variable $X$.

**EXPLORE 3.4**

From the outcomes of mean and variance with Xing's and Quenby's dice and the probability distribution for $W$, we note that $E(2X) = E(X_1 + X_2)$, whereas $\text{Var}(2X) \neq \text{Var}(X_1 + X_2)$. In similar situations, will it always be true that the expectation is the same but the variance is different? Explore the outcomes with different random variables.

## EXERCISE 3B

1 The independent random variables $X$ and $Y$ are such that $X \sim B(20, 0.3)$ and $Y \sim B(12, 0.4)$. Find:

  a $E(2X - Y)$  b $\text{Var}(2X - Y)$

2 The independent random variables $X$ and $Y$ are such that $X \sim B(8, 0.25)$ and $Y \sim B(5, 0.6)$. Find:

  a $E(6X + 2Y)$  b $\text{Var}(3X - 2Y + 1)$

3 The independent random variables $X$ and $Y$ have means 5 and 6, and variances 0.2 and 0.3, respectively. Calculate the mean and variance of:

a $3X + 4Y$ b $2X - 3Y$

4 The independent random variables $X$ and $Y$ have standard deviations 4 and 7, respectively. Calculate the standard deviation of:

a $2X + 3Y$ b $X - 2Y$

5 The independent random variables $A$, $B$ and $C$ have means 4, 9, and 7, respectively, and variances 1, 2 and 1.3, respectively. Find:

a $E(A + B + C)$ b $Var(A + B + C)$

c $E(3A + B - 2C)$ d $Var(3A + B - 2C)$

6 Given that the random variable $X$ has probability distribution as given in the table, find:

| $x$ | 2 | 3 | 5 |
|---|---|---|---|
| $P(X = x)$ | 0.6 | 0.1 | 0.3 |

a $E(X)$ and $Var(X)$ b $E(2X)$ and $Var(2X)$

c the expectation of two independent observations of $X$, $E(X_1 + X_2)$

d the variance of two independent observations of $X$, $Var(X_1 + X_2)$.

7 The random variable $S$ is the score when a fair ordinary die is rolled once, and the random variable $H$ is the number of heads obtained when two unbiased coins are tossed once. Find the mean, variance and standard deviation of the random variable $T$, where $T = 4S - 3H$.



## 3.3 Working with normal distributions

Physical quantities that are the sum of normally distributed independent variables are themselves normally distributed. For example, a cricket bat is constructed by joining a blade to a handle, each of which has lengths that are normally distributed, and the resultant lengths of cricket bats are themselves normally distributed.

**KEY POINT 3.6**

If a continuous random variable $X$ has a normal distribution, then $aX + b$, where $a$ and $b$ are constants, also has a normal distribution.

If continuous random variables $X$ and $Y$ have independent normal distributions, then $aX + bY$, where $a$ and $b$ are constants, has a normal distribution.

**REWIND**

In the Probability & Statistics 1 Coursebook, Chapter 8, you learnt that many naturally occurring phenomena or quantities would be preferable and follow an approximately normal distribution.

You also learnt that you could standardise a normal distribution $X \sim N(\mu, \sigma^2)$ to $Z \sim N(0, 1)$. This generalisation gives $z = \frac{x - \mu}{\sigma}$, which rearranges to $x = \sigma z + \mu$ and, from Key point 3.3, this is a linear combination of a random variable, and both $Z$ and $X$ follow a normal distribution.

**WORKED EXAMPLE 3.6**

The masses, in kilograms, of small bags of rice and large bags of rice are denoted by $X$ and $Y$, respectively, where $X \sim \mathrm{N}(2.1, 0.2^2)$ and $Y \sim \mathrm{N}(6.6, 0.4^2)$. Find the probability that the mass of a randomly chosen large bag is greater than three times the mass of a randomly chosen small bag.

**Answer**

$Y - 3X \sim \mathrm{N}(6.6 - 3(2.1), 0.4^2 + 3^2 0.2^2)$

$Y - 3X \sim \mathrm{N}(0.3, 0.52)$

If $Y > 3X$, then $Y - 3X > 0$. So we need $Y - 3X$ to be greater than zero if the large bag is greater than three smaller bags.

Use Key point 3.6 to find the mean and variance of $Y - 3X$.

$$P(Y - 3X > 0) = 1 - \Phi\left(\frac{0 - 0.3}{\sqrt{0.52}}\right)$$

$$= 1 - \Phi(-0.416)$$

$$= 0.6613$$

$$= 0.661 \text{ (to 3 significant figures)}$$

**WORKED EXAMPLE 3.7**

The lifetime, in months, of Conti batteries and Thrift batteries have the independent distributions $C \sim \mathrm{N}(25, 1.1^2)$ and $T \sim \mathrm{N}(7, 2.3^2)$, respectively.

**a** **i** What is the expected total lifetime and variance of four randomly chosen Thrift batteries?

**ii** Find the probability that the total lifetime of four randomly chosen Thrift batteries is longer than 30 months.

**b** What is the probability that the lifetime of a randomly chosen Conti battery is at least four times that of a Thrift battery?



**Answer**

Let $S$ represent the sum of the lifetimes of four independent Thrift batteries.

In this question you have the sum of four independent observations of the random variable $T$; look back at the Tip included in the section 'When is $2X \neq X + X$?'.

**a** **i** $\mathrm{E}(S) = 7 + 7 + 7 + 7 = 28$

$\mathrm{Var}(S) = 2.3^2 + 2.3^2 + 2.3^2 + 2.3^2 = 21.16$

**ii** $1 - \Phi\left(\frac{30 - 28}{\sqrt{21.16}}\right) = 1 - \Phi(0.435)$

Remember to use normal tables.

$$= 1 - 0.6682$$

$$= 0.332 \text{ (to 3 s.f.)}$$

**b** $\mathrm{E}(C - 4T) = \mathrm{E}(C) - 4\mathrm{E}(T) = 25 - 4 \times 7 = -3$

In this question you have four times the size the random variable $T$; look back at the Tip included in the section 'When is $2X \neq X + X$?'.

$$\mathrm{Var}(C - 4T) = \mathrm{Var}(C) + 4^2\mathrm{Var}(T)$$

$$= 1.1^2 + 16 \times 2.3^2$$

$$= 85.9$$

$$P(C - 4T > 0) = 1 - \Phi\left(\frac{0 - (-3)}{\sqrt{85.9}}\right)$$

$$= 1 - \Phi(0.324)$$

$$= 1 - 0.6270$$

$$= 0.373$$

**WORKED EXAMPLE 3.8**

A melamine worktop is made by joining sheets of melamine and chipboard. The distribution of the thickness of the chipboard, $C$, in mm, is $C \sim N(37, 0.3^2)$ and the distribution of the thickness of the melamine, $M$, in mm, is $M \sim N(1, 0.01^2)$. Assuming the distributions are independent, find the mean and variance of the thickness of the worktop if the melamine sheet is added to:

**a** just the top surface of the chipboard

**b** both the top and bottom surfaces of the chipboard.

**Answer**

**a** Mean $= 37 + 1 = 38$

Use the result from Key point 3.5.

Variance $= 0.3^2 + 0.01^2 = 0.0901$

**b** Mean $= 37 + 1 + 1 = 39$

Note that there are two layers of melamine and these are independent.

Variance $= 0.3^2 + 0.01^2 + 0.01^2 = 0.0902$

**WORKED EXAMPLE 3.9**

A gift package contains a set of three bars of soap. The mass, in grams, of each bar of soap is given by $S \sim N(50, 4^2)$ and the packaging by $T \sim N(90, 5^2)$. The gift package can be posted at a cheap rate if the mass is under 250 g. Assuming that $S$ and $T$ are independent, find the probability that a randomly chosen gift package can be posted at the cheap rate.

**Answer**

$3S + T \sim N(240, 13^2)$

The gift package comprises three bars of soap, $3S$, and the packaging, $T$. So $3S + T$ is the random variable for the total mass of the gift package, mean $= 150 + 90$.

$$\text{Var}(3S + T) = 3^2\text{Var}(S) + \text{Var}(T)$$
$$= 144 + 25$$
$$= 169$$

To complete the solution, find $P(3S + T < 250)$.

$$\Phi\left(\frac{250 - 240}{13}\right) = \Phi(0.769) = 0.779$$



**EXERCISE 3C**

1 The distributions of the independent random variables $A$, $B$ and $C$ are $A \sim N(10, 4)$, $B \sim N(28, 25)$ and $C \sim N(15, 3^2)$. Write down the distributions of:

**a** $A + 2B$ **b** $2B - A$ **c** $A + B + C$

2 The distributions of the independent random variables $X$ and $Y$ are $X \sim N(24, 2^2)$ and $Y \sim N(25, 3^2)$.

**a** Write down the distribution of $2X - Y$.

**b** Find $P(2X - Y < 28)$.

3 The distributions of the independent random variables $X$ and $Y$ are $X \sim N(80, 8^2)$ and $Y \sim N(48, 6^2)$. Find:

**a** $P(X + Y > 130)$ **b** $P(X - Y < 40)$

PS 4 The masses, in kilograms, of small bags of flour and large bags of flour are denoted by $X$ and $Y$, respectively, where $X \sim N(0.51, 0.1^2)$ and $Y \sim N(1.1, 0.25^2)$. Find the probability that the mass of a randomly chosen large bag is less than two times the mass of a randomly chosen small bag.

PS 5 The distribution of the mass, in grams, of honey $H$ in a jar is given by $H \sim N(465, 25)$. The distribution of the mass, in grams, of the jar $J$ is given by $J \sim N(120, 16)$. Find the expected total mass of 12 of these jars of honey and the standard deviation of the mass, assuming the masses of jars and honey are independent and given in the same units.

PS 6 The time taken for an athlete to run 100 m is normally distributed. The mean times, in seconds, for four athletes in a $4 \times 100$ m relay team are 22.2, 24.8, 23.6 and 20.4, and the standard deviations of times are 0.35, 0.24, 0.38 and 0.16, respectively. The times for each athlete are independent. Find the probability that in a randomly chosen $4 \times 100$ m relay race, their time is less than 90 seconds.

PS 7 The masses of cupcakes produced at a bakery are independent and modelled as being normally distributed with mean 120 g and standard deviation 5 g. Find the probability that six cupcakes have a total mass between 700 g and 725 g.

M 8 A paper mill produces sheets of paper, each of a constant thickness. The thickness of the paper, in mm, is defined as the random variable $T$, where $T \sim N(0.1, 0.00005)$.

a Write down the distribution of the total thickness of 16 randomly chosen sheets of paper.

b A single sheet of paper is folded four times so that the result is 16 times as thick as a single sheet. Write down the distribution of the total thickness.



## 3.4 Linear combinations of Poisson distributions

The staff of a lion and tiger rescue centre found that they rescued, on average, five lions and three tigers per week. The number of lions and tigers rescued can be modelled by Poisson distributions, $L \sim Po(5)$ and $T \sim Po(3)$, respectively. In one week, only two animals were rescued. The possible combinations are no lions and two tigers or one lion and one tiger or two lions and no tigers. Here is the calculation to find the probability that two animals are rescued in one week:

$$P(L = 0) \times P(T = 2) + P(L = 1) \times P(T = 1) + P(L = 2) \times P(T = 0)$$

$$= e^{-5}\frac{5^0}{0!} \times e^{-3}\frac{3^2}{2!} + e^{-5}\frac{5^1}{1!} \times e^{-3}\frac{3^1}{1!} + e^{-5}\frac{5^2}{2!} \times e^{-3}\frac{3^0}{0!}$$

$$= (0.00674 \times 0.224) + (0.0337 \times 0.149) + (0.0842 \times 0.0498)$$

$$= 0.00151 + 0.00502 + 0.00419 = 0.0107$$

It is not hard to see that this is a time-consuming calculation.

Suppose, instead, we combined the data from the rescue centre and considered the number of lions and tigers rescued each week. Let the random variable $A$ be the number of lions and tigers rescued each week; then $A \sim Po(8)$.

$$P(A = 2) = e^{-8}\frac{8^2}{2!} = 0.0107$$

This is the same result as that reached by the rather longer calculation working with possible combinations.

If you work out the two methods using different numbers of lions and tigers rescued, you will find that the same result holds. Here is an alternative way to write this: if $L \sim Po(5)$ and $T \sim Po(3)$, then $L + T \sim Po(8)$.

**KEY POINT 3.7**

If $X$ and $Y$ have independent Poisson distributions, then $X + Y$ has a Poisson distribution.

If $X \sim \text{Po}(\lambda)$ and $Y \sim \text{Po}(\mu)$, then $X + Y \sim \text{Po}(\lambda + \mu)$.

This result can be extended to any number of independent Poisson distributions.

**EXPLORE 3.5**

Explore this result for more than two independent Poisson distributions. For example, if the average number of cheetahs rescued each week can be modelled by a Poisson distribution $C \sim \text{Po}(2)$, and $L \sim \text{Po}(5)$ and $T \sim \text{Po}(3)$ as before, work out the probability of any two of cheetahs, lions and tigers being rescued in one week and compare this to $\text{P}(S = 2)$, where $S \sim \text{Po}(10)$.

**WORKED EXAMPLE 3.10**

Two brothers, Josh and Reuben, keep in touch with their parents by text. The numbers of text messages sent by Josh and Reuben are independent. The number of text messages sent each week by Josh has a Poisson distribution with mean 3.2. The number of text messages sent by Reuben each week has a Poisson distribution with mean 2.5.

**a** Write down the distribution for $T$, where $T$ is a random variable defined as 'the number of text messages sent each week by Josh and Reuben'.

**b** Calculate $\text{P}(T > 5)$.

**Answer**

**a** $T \sim \text{Po}(5.7)$

$3.2 + 2.5 = 5.7$. Write the distribution fully, do not write just 'Poisson' or '$\lambda = 5.7$'.

**b** $\text{P}(T > 5) = 1 - \text{P}(T \leqslant 4)$

$$= 1 - \text{e}^{-5.7}\left(1 + 5.7 + \frac{5.7^2}{2!} + \frac{5.7^3}{3!} + \frac{5.7^4}{4!}\right)$$

$$= 1 - 0.327 = 0.673$$

Remember, the probability of 'greater than' is the same as the probability '1 – less than or equal to'.

**WORKED EXAMPLE 3.11**

The independent random variables $X$ and $Y$ are given by $X \sim \text{Po}(2.4)$ and $Y \sim \text{Po}(3.6)$. The random variable $T = 2X - Y$. Work out the expectation and variance of $T$. Explain why the random variable $T$ does not follow a Poisson distribution.

**Answer**

$\text{E}(T) = \text{E}(2X - Y) = 2\text{E}(X) - \text{E}(Y) = 2 \times 2.4 - 3.6 = 1.2$

$\text{Var}(T) = \text{Var}(2X - Y) = 2^2\text{Var}(X) + (-1)^2\text{Var}(Y)$

$= 4 \times 2.4 + 3.6 = 13.2$

The random variable $T$ does not follow a Poisson distribution because the mean and variance are not equal.

For a random variable to follow a Poisson distribution the mean and variance have to be equal.

Note that this example shows that when you find the mean and variance of multiples of independent random Poisson variables, the resulting distribution does not follow a Poisson distribution.

## EXERCISE 3D

1 The independent random variables $X$ and $Y$ are such that $X \sim \text{Po}(4.2)$ and $Y \sim \text{Po}(6.3)$. Find the expectation and variance of the following.

a $2X - Y$ b $X + Y$ c $X + 3Y$

d Which of parts **a**, **b** and **c** has a Poisson distribution?

2 The independent random variables $X$ and $Y$ are such that $X \sim \text{Po}(7)$ and $Y \sim \text{Po}(5)$. Find:

a $\text{E}(X + Y)$ b $\text{P}(X + Y = 2)$

3 The independent random variables $X$ and $Y$ are such that $X \sim \text{Po}(3.5)$ and $Y \sim \text{Po}(4)$. Find $\text{P}(X + Y \leqslant 1)$.

4 The independent random variables $X$ and $Y$ are such that $X \sim \text{Po}(3.2)$ and $Y \sim \text{Po}(1.3)$. Work out:

a $\text{P}(X + Y = 6)$ b $\text{P}(X + Y < 1)$

5 The independent random variables $X$ and $Y$ are such that $X \sim \text{Po}(2)$ and $Y \sim \text{Po}(1)$. State the distribution of $T$, where $T = X + Y$. Find $\text{P}(2 \leqslant T < 4)$.

6 The independent random variables $D$ and $E$ are such that $D \sim \text{Po}(0.7)$ and $E \sim \text{Po}(1.3)$. Work out:

a $\text{P}(D + E = 5)$ b $\text{P}(D + E \geqslant 2)$

7 The independent random variables $A$, $B$ and $C$ are such that $A \sim \text{Po}(2)$, $B \sim \text{Po}(1.2)$ and $C \sim \text{Po}(0.5)$.

a State the distribution of $T$, where $T = A + B + C$.

b Work out:

i $\text{P}(T = 4)$ ii $\text{P}(T < 3)$



PS 8 The number of cars passing a fixed point on a road can be modelled as a random variable having a Poisson distribution with mean two cars per hour. The number of lorries passing the same point on the road can be modelled as a random variable having a Poisson distribution with mean three lorries per hour. Assuming that the random variables are independent, find the probability that in a randomly chosen hour a total of exactly four of these vehicles pass that point on the road.

PS 9 The number of letters and parcels received per weekday by a business are independent. On average, the business receives three letters and one parcel per weekday. State an assumption needed to model the number of items received using a Poisson distribution, and find:

a the probability that the total number of items received on a randomly chosen weekday exceeds five

b the probability that in a randomly chosen working week of five days, the business receives more than 20 but fewer than 24 items.

M 10 The numbers of goals scored in a football match by teams $X$ and $Y$ can be modelled by Poisson distributions with means 0.7 and 1.2, respectively. Find the probability that in a match between teams $X$ and $Y$ there are, at most, two goals scored, stating any assumptions you make. Comment on whether your assumptions are reasonable in this situation.

## Checklist of learning and understanding

For a random variable $X$ and constants $a$ and $b$:

- $\mathrm{E}(aX + b) = a\mathrm{E}(X) + b$
- $\mathrm{Var}(aX + b) = a^2\mathrm{Var}(X)$

For independent random variables $X$ and $Y$ and constants $a$ and $b$:

- $\mathrm{E}(aX + bY) = a\mathrm{E}(X) + b\mathrm{E}(Y)$
- $\mathrm{Var}(aX + bY) = a^2\mathrm{Var}(X) + b^2\mathrm{Var}(Y)$

If $X$ has a normal distribution, then so does $aX + b$.

If $X$ and $Y$ have independent normal distributions, then $aX + bY$ has a normal distribution.

If independent random variables $X$ and $Y$ have Poisson distributions, then $X + Y$ has a Poisson distribution.

## END-OF-CHAPTER REVIEW EXERCISE 3

1 Over a long period, it has been found that the average temperature in an area of California is 17.6 °C. The standard deviation in temperature is 4 °C. Using the conversion F = 1.8C + 32, where F is the temperature in degrees Fahrenheit (°F) and C is the temperature in degrees Celsius (°C), work out the average temperature and standard deviation in temperature in degrees Fahrenheit. **[3]**

2 The independent random variables $X$ and $Y$ have the distributions $N(6.5, 14)$ and $N(7.4, 15)$, respectively. Find $P(3X - Y < 20)$. **[5]**

*Cambridge International AS & A Level Mathematics 9709 Paper 73 Q2 June 2012*

3 Large eggs have mean mass 68 g and standard deviation 1.7 g. The packaging for a box of six eggs has mean mass 36 g and standard deviation 4 g. Given that the masses of eggs are independent, find the probability that a randomly chosen box of six eggs has mass greater than 450 g. **[5]**

4 Porridge oats are packed in two different-sized bags, large and small. The masses, in grams, of the large and small bags are given by $L$ and $S$, respectively, where $L \sim N(1010, 10^2)$ and $S \sim N(504, 8^2)$. Assuming that $L$ and $S$ are independent, find the probability that the mass of a randomly chosen large bag is more than twice the mass of a randomly chosen small bag. **[6]**

5 In an 'eight' rowing boat there are eight rowers and a coxswain, who gives instructions to the rowers. The distribution of masses, in kg, of rowers is given by $M \sim N(92.5,\ 1.8^2)$. The distribution of the mass, in kg, of the coxswain is given by $C \sim N(57, 0.6^2)$. Assuming that the masses of the rowers are independent, find the probability that in a randomly chosen 'eight' rowing boat the total mass of the rowers and coxswain is less than 800 kg. **[5]**

6 A men's triathlon consists of three parts: swimming, cycling and running. Competitors' times, in minutes, for the three parts can be modelled by three independent normal variables with means 34.0, 87.1 and 56.9, and standard deviations 3.2, 4.1 and 3.8, respectively. For each competitor, the total of his three times is called the race time. Find the probability that the mean race time of a random sample of 15 competitors is less than 175 minutes. **[5]**

*Cambridge International AS & A Level Mathematics 9709 Paper 73 Q3 November 2016*

7 Kieran and Andreas are long-jumpers. They model the lengths, in metres, that they jump by the independent random variables $K \sim N(5.64, 0.0576)$ and $A \sim N(4.97, 0.0441)$, respectively. They each make a jump and measure the length. Find the probability that:

i the sum of the lengths of their jumps is less than 11 m **[4]**

ii Kieran jumps more than 1.2 times as far as Andreas. **[6]**

*Cambridge International AS & A Level Mathematics 9709 Paper 71 Q7 November 2013*

8 The mean and variance of the random variable $X$ are 5.8 and 3.1, respectively. The random variable $S$ is the sum of three independent values of $X$. The independent random variable $T$ is defined by $T = 3X + 2$.

i Find the variance of $S$. **[1]**

ii Find the variance of $T$. **[1]**

iii Find the mean and variance of $S - T$. **[3]**

*Cambridge International AS & A Level Mathematics 9709 Paper 73 Q1 June 2013*

9 The cost of hiring a bicycle consists of a fixed charge of 500 cents together with a charge of 3 cents per minute. The number of minutes for which people hire a bicycle has mean 142 and standard deviation 35.

i Find the mean and standard deviation of the amount people pay when hiring a bicycle. **[3]**

ii Six people hire bicycles independently. Find the mean and standard deviation of the total amount paid by all six people. **[3]**

*Cambridge International AS & A Level Mathematics 9709 Paper 72 Q3 November 2012*

10 The masses, in milligrams, of three minerals found in 1 tonne of a certain kind of rock are modelled by three independent random variables $P$, $Q$ and $R$, where $P \sim \mathrm{N}(46, 19^2)$, $Q \sim \mathrm{N}(53, 23^2)$ and $R \sim \mathrm{N}(25, 10^2)$. The total value of the minerals found in 1 tonne of rock is modelled by the random variable $V$, where $V = P + Q + 2R$. Use the model to find the probability of finding minerals with a value of at least 93 in a randomly chosen tonne of rock. **[7]**

*Cambridge International AS & A Level Mathematics 9709 Paper 73 Q4 November 2010*

11 The weekly distance, in kilometres, driven by Mr Parry has a normal distribution with mean 512 and standard deviation 62. Independently, the weekly distance, in kilometres, driven by Mrs Parry has a normal distribution with mean 89 and standard deviation 7.4.

i Find the probability that, in a randomly chosen week, Mr Parry drives more than five times as far as Mrs Parry. **[5]**

ii Find the mean and standard deviation of the total of the weekly distances, in miles, driven by Mr Parry and Mrs Parry. Use the approximation 8 kilometres = 5 miles. **[3]**

*Cambridge International AS & A Level Mathematics 9709 Paper 71 Q4 June 2010*

12 i Random variables $Y$ and $X$ are related by $Y = a + bX$, where $a$ and $b$ are constants and $b > 0$. The standard deviation of $Y$ is twice the standard deviation of $X$. The mean of $Y$ is 7.92 and is 0.8 more than the mean of $X$. Find the values of $a$ and $b$. **[3]**

ii Random variables $R$ and $S$ are such that $R \sim \mathrm{N}(\mu, 2^2)$ and $S \sim \mathrm{N}(2\mu, 3^2)$. It is given that $\mathrm{P}(R + S > 1) = 0.9$.

a Find $\mu$. **[4]**

b Hence, find $\mathrm{P}(S > R)$. **[3]**

*Cambridge International AS & A Level Mathematics 9709 Paper 72 Q7 November 2009*

## CROSS-TOPIC REVIEW EXERCISE 1

1 In the manufacture of a certain material, faults occur independently and at random at an average of 0.14 per 1 m$^2$. To make a particular design of shirt, 2.5 m$^2$ of this material is required.

a Find the probability that in a randomly selected 2.5 m$^2$ area of the material there is, at most, one fault. **[3]**

The material is going to be used to make a batch of ten shirts, each requiring 2.5 m$^2$ of material.

b Find the probability that in exactly three of these ten shirts there will be exactly two faults. **[3]**

The buttons for these shirts are produced by a company that claims only 1% of the buttons it produces are defective.

c From a randomly selected batch of 120 of these buttons, three are found to be defective. Test the claim at the 2% significance level. **[4]**

2 The maximum load an elevator can carry is 600 kg. The masses of men are normally distributed with mean 80 kg and standard deviation 9 kg. The masses of women are normally distributed with mean 65 kg and standard deviation 6 kg. Assuming the masses of men and women are independent, find the probability that the elevator will not be overloaded by a group of six men and two women. **[4]**

3 The number of calls received per 5-minute period at a large call centre has a Poisson distribution with mean $\lambda$, where $\lambda > 30$. If more than 55 calls are received in a 5-minute period, the call centre is overloaded. It has been found that the probability of being overloaded during a randomly chosen 5-minute period is 0.01. Use the normal approximation to the Poisson distribution to obtain a quadratic equation in $\sqrt{\lambda}$ and, hence, find the value of $\lambda$. **[5]**



*Cambridge International AS & A Level Mathematics 9709 Paper 73 Q2 November 2015*

4 People arrive at a checkout in a store at random, and at a constant mean rate of 0.7 per minute. Find the probability that:

i exactly 3 people arrive at the checkout during a 5-minute period **[2]**

ii at least 30 people arrive at the checkout during a 1-hour period. **[4]**

People arrive independently at another checkout in the store at random, and at a constant mean rate of 0.5 per minute.

iii Find the probability that a total of more than 3 people arrive at this pair of checkouts during a 2-minute period. **[4]**

*Cambridge International AS & A Level Mathematics 9709 Paper 73 Q6 June 2015*

5 The number of accidents on a certain road has a Poisson distribution with mean 3.1 per 12-week period.

i Find the probability that there will be exactly 4 accidents during an 18-week period. **[3]**

Following the building of a new junction on this road, an officer wishes to determine whether the number of accidents per week has decreased. He chooses 15 weeks at random and notes the number of accidents. If there are fewer than 3 accidents altogether he will conclude that the number of accidents per week has decreased. He assumes that a Poisson distribution still applies.

ii Find the probability of a Type I error. **[3]**

iii Given that the mean number of accidents per week is now 0.1, find the probability of a Type II error. **[3]**

**iv** Given that there were 2 accidents during the 15 weeks, explain why it is impossible for the officer to make a Type II error. **[1]**

*Cambridge International AS & A Level Mathematics 9709 Paper 72 Q6 November 2014*

**6** In an examination, the marks in the theory paper and the marks in the practical paper are denoted by the random variables $X$ and $Y$, respectively, where $X \sim N(57, 13)$ and $Y \sim N(28, 5)$. You may assume that each candidate's marks in the two papers are independent. The final score of each candidate is found by calculating $X + 2.5Y$. A candidate is chosen at random. Without using a continuity correction, find the probability that this candidate:

**i** has a final score that is greater than 140 **[5]**

**ii** obtains at least 20 more marks in the theory paper than in the practical paper. **[5]**

*Cambridge International AS & A Level Mathematics 9709 Paper 73 Q8 June 2014*

**7** At the last election, 70% of people in Apoli supported the president. Luigi believes that the same proportion support the president now. Maria believes that the proportion who support the president now is 35%. In order to test who is right, they agree on a hypothesis test, taking Luigi's belief as the null hypothesis. They will ask 6 people from Apoli, chosen at random, and if more than 3 support the president they will accept Luigi's belief.

**i** Calculate the probability of a Type I error. **[3]**

**ii** If Maria's belief is true, calculate the probability of a Type II error. **[3]**

**iii** In fact 2 of the 6 people say that they support the president. State which error, Type I or Type II, might be made. Explain your answer. **[2]**

*Cambridge International AS & A Level Mathematics 9709 Paper 71 Q6 November 2013*

**8** The independent random variables $X$ and $Y$ have the distributions Po(2) and Po(3), respectively.

**i** Given that $X + Y = 5$, find the probability that $X = 1$ and $Y = 4$. **[4]**

**ii** Given that $P(X = r) = \frac{2}{3} P(X = 0)$, show that $3 \times 2^{r-1} = r!$ and verify that $r = 4$ satisfies this equation. **[2]**

*Cambridge International AS & A Level Mathematics 9709 Paper 73 Q4 June 2013*

**9** A random variable $X$ has the distribution Po(1.6).

**i** The random variable $R$ is the sum of three independent values of $X$. Find $P(R < 4)$. **[3]**

**ii** The random variable $S$ is the sum of $n$ independent values of $X$. It is given that $P(S = 4) = \frac{16}{3} \times P(S = 2)$. Find $n$. **[4]**

**iii** The random variable $T$ is the sum of 40 independent values of $X$. Find $P(T > 75)$. **[4]**

*Cambridge International AS & A Level Mathematics 9709 Paper 72 Q7 November 2012*

**10** The number of lions seen per day during a standard safari has the distribution Po(0.8). The number of lions seen per day during an off-road safari has the distribution Po(2.7). The two distributions are independent.

**i** Susan goes on a standard safari for one day. Find the probability that she sees at least 2 lions. **[2]**

**ii** Deena goes on a standard safari for 3 days and then on an off-road safari for 2 days. Find the probability that she sees a total of fewer than 5 lions. **[3]**

**iii** Khaled goes on a standard safari for $n$ days, where $n$ is an integer. He wants to ensure that his chance of not seeing any lions is less than 10%. Find the smallest possible value of $n$. **[3]**

*Cambridge International AS & A Level Mathematics 9709 Paper 73 Q4 June 2012*

**11** An airline knows that some people who have bought tickets may not arrive for the flight. The airline therefore sells more tickets than the number of seats that are available. For one flight there are 210 seats available and 213 people have bought tickets. The probability of any person who has bought a ticket not arriving for the flight is $\frac{1}{50}$.

**i** By considering the number of people who do **not** arrive for the flight, use a suitable approximation to calculate the probability that more people will arrive than there are seats available. **[4]**

Independently, on another flight for which 135 people have bought tickets, the probability of any person not arriving is $\frac{1}{75}$.

**ii** Calculate the probability that, for both these flights, the total number of people who do not arrive is 5. **[3]**

*Cambridge International AS & A Level Mathematics 9709 Paper 72 Q3 November 2009*

**12** Every month Susan enters a particular lottery. The lottery company states that the probability, $p$, of winning a prize is 0.0017 each month. Susan thinks that the probability of winning is higher than this, and carries out a test based on her 12 lottery results in a one-year period. She accepts the null hypothesis $p = 0.0017$ if she has no wins in the year and accepts the alternative hypothesis $p > 0.0017$ if she wins a prize in at least one of the 12 months.



**i** Find the probability of the test resulting in a Type I error. **[2]**

**ii** If in fact the probability of winning a prize each month is 0.0024, find the probability of the test resulting in a Type II error. **[3]**

**iii** Use a suitable approximation, with $p = 0.0024$, to find the probability that in a period of 10 years Susan wins a prize exactly twice. **[3]**

*Cambridge International AS & A Level Mathematics 9709 Paper 7 Q5 November 2008*

**13** The lengths of red pencils are normally distributed with mean 6.5 cm and standard deviation 0.23 cm.

**i** Two red pencils are chosen at random. Find the probability that their total length is greater than 12.5 cm. **[3]**

The lengths of black pencils are normally distributed with mean 11.3 cm and standard deviation 0.46 cm.

**ii** Find the probability that the total length of 3 red pencils is more than 6.7 cm greater than the length of 1 black pencil. **[4]**

*Cambridge International AS & A Level Mathematics 9709 Paper 7 Q3 June 2008*

**14** When a guitar is played regularly, a string breaks on average once every 15 months. Broken strings occur at random times and independently of each other.

**i** Show that the mean number of broken strings in a 5-year period is 4. **[1]**

A guitar is fitted with a new type of string which, it is claimed, breaks less frequently. The number of broken strings of the new type was noted after a period of 5 years.

**ii** The mean number of broken strings of the new type in a 5-year period is denoted by $\lambda$. Find the rejection region for a test at the 10% significance level when the null hypothesis $\lambda = 4$ is tested against the alternative hypothesis $\lambda < 4$. **[4]**

**iii** Hence, calculate the probability of making a Type I error. **[1]**

The number of broken guitar strings of the new type, in a 5-year period, was in fact 1.

**iv** State, with a reason, whether there is evidence at the 10% significance level that guitar strings of the new type break less frequently. **[2]**

*Cambridge International AS & A Level Mathematics 9709 Paper 7 Q5 June 2008*

# Chapter 4
# Continuous random variables

**In this chapter you will learn how to:**

- understand the concept of a continuous random variable
- recall and use the properties of a probability density function
- calculate the mean and variance of a continuous distribution
- find the median and other percentiles of a distribution
- solve problems involving probabilities.

**PREREQUISITE KNOWLEDGE**

| Where it comes from | What you should be able to do | Check your skills |
|---|---|---|
| Pure Mathematics 1, Chapter 9<br>Pure Mathematics 3, Chapter 8 | Evaluate definite integrals. | **1** Evaluate $\int_1^3 2x^3 - 5x + 1 \, dx$.<br>**2** Evaluate $\int_{0.2}^{0.3} 8 \, dx$.<br>**3** Evaluate $\int_4^9 \frac{5}{\sqrt{x}} \, dx$.<br>**4** Evaluate $\int_2^5 e^{-0.4x} \, dx$.<br>**5** Find $m$, where $\int_0^m 6x - 1 \, dx = 0.25$. |

## Why do we study continuous random variables?

In Probability & Statistics Coursebook 1, you learnt how to use the normal distribution as a model for continuous random variables. This is the most commonly encountered continuous random variable since many naturally occurring phenomena, such as length of index finger or time of natural sleep or the speed of cars passing a particular point on a road, consist of continuous data that can be described using a normal distribution. However, you cannot use a normal distribution to describe all continuous random variables; for example, the length of time cars stay in a long-term car park or the rate of decay of a radioactive element, because they are strongly skewed. This chapter looks at other ways to model continuous random variables.



**EXPLORE 4.1**

Age can be modelled as a discrete or **continuous random variable**. Except for very young children, age is usually given as a whole number of years; counting the number of years means age is a discrete random variable. If, for the purposes of a study, you need to know an exact age, then you are still measuring age and it can still be considered a continuous random variable. Discuss why the way in which you define the random variable is important. Can you give other similar examples?

## 4.1 Introduction to continuous random variables

**REWIND**

In Chapter 1 of Probability & Statistics Coursebook 1, the histogram was introduced as an appropriate way to display continuous data. A crucial property of the histogram is that the area of bars represents frequencies, so that the heights of bars show frequency densities.

Let us consider an example. In Chapter 2 we looked at the number of patients arriving at a minor injuries clinic. In the survey, data were also collected on the time each of the 425 patients waited to see a doctor or nurse; 257 patients waited up to 30 minutes, 132 patients waited for 30 minutes to 60 minutes and 36 patients waited for 60 minutes to 120 minutes. This continuous data, described in three class intervals, can be represented in a histogram, as shown in the following diagram, which plots the waiting time, in minutes, in three class intervals.

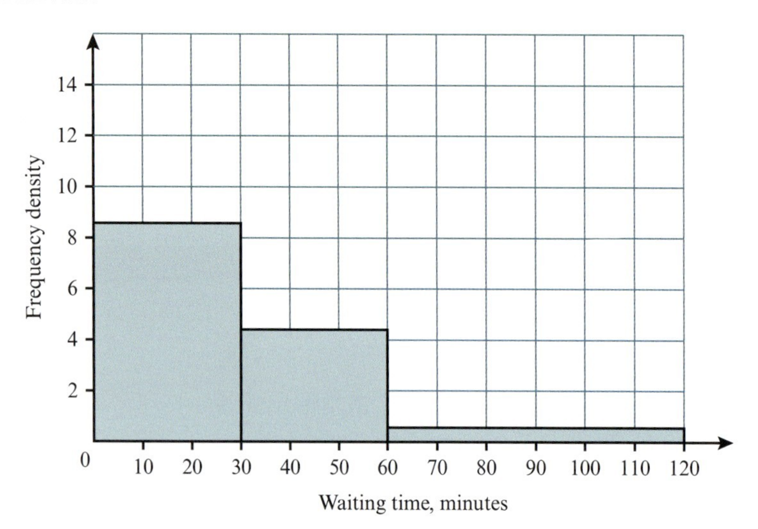




The same data, collated in eight smaller class intervals, are shown in the following table.

| **Waiting time, *t* minutes** | $0 \leqslant t < 5$ | $5 \leqslant t < 10$ | $10 \leqslant t < 20$ | $20 \leqslant t < 30$ | $30 \leqslant t < 45$ | $45 \leqslant t < 60$ | $60 \leqslant t < 90$ | $90 \leqslant t < 120$ |
|---|---|---|---|---|---|---|---|---|
| **Frequency** | 14 | 30 | 93 | 120 | 95 | 37 | 28 | 8 |

The data can also be plotted as a histogram. The result is shown in the following diagram, which plots the waiting time, in minutes, in eight class intervals.

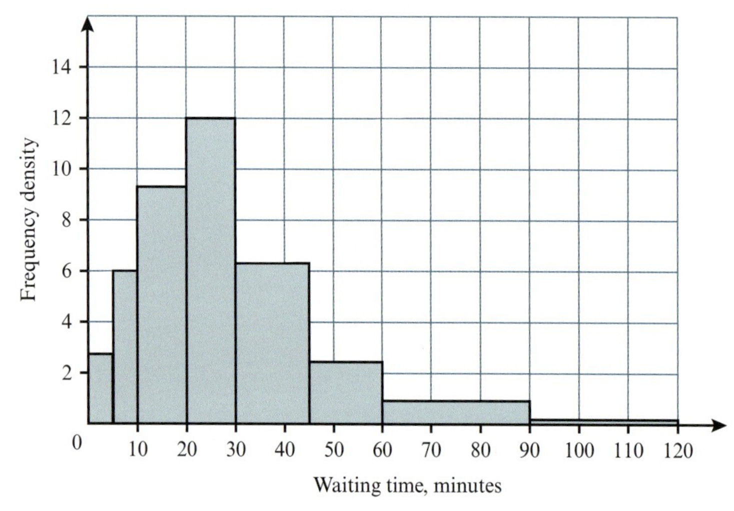


The same data could be collated in narrower class intervals. The results are shown in the following table and its corresponding histogram, which plots the waiting time, in minutes, in 17 class intervals.

| Waiting time, $t$ minutes | $0 \leqslant t < 5$ | $5 \leqslant t < 10$ | $10 \leqslant t < 15$ | $15 \leqslant t < 20$ | $20 \leqslant t < 25$ | $25 \leqslant t < 30$ | $30 \leqslant t < 35$ | $35 \leqslant t < 40$ | $40 \leqslant t < 45$ |
|---|---|---|---|---|---|---|---|---|---|
| **Frequency** | 14 | 30 | 42 | 51 | 66 | 54 | 42 | 32 | 21 |
| **Waiting time, $t$ minutes** | $45 \leqslant t < 50$ | $50 \leqslant t < 55$ | $55 \leqslant t < 60$ | $60 \leqslant t < 70$ | $70 \leqslant t < 80$ | $80 \leqslant t < 90$ | $90 \leqslant t < 105$ | $105 \leqslant t < 120$ | |
| **Frequency** | 17 | 12 | 8 | 11 | 9 | 8 | 6 | 2 | |

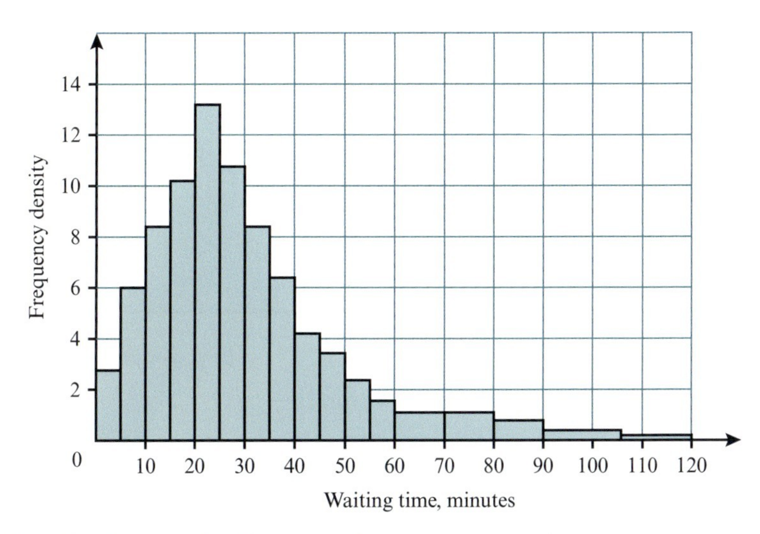


We can see that the shape of the histogram begins to approximate a curve.

If we use the probabilities of waiting times to calculate relative frequency density, then the graph will have the same shape, and the vertical axis scale is probability density, as shown in the following diagram.

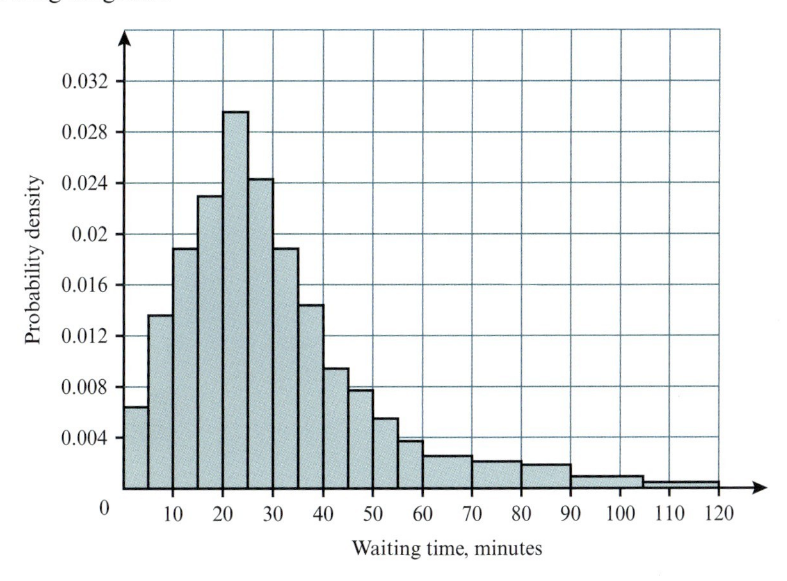


**REWIND**

Frequency density = frequency ÷ class width

Probability density = frequency density ÷ total frequency

For example, if waiting time is 15–20 minutes:

Frequency density $= 51 \div 5 = 10.2$ and probability density $= 10.2 \div 425 = 0.024$.

**REWIND**

In Chapter 8 of Probability & Statistics Coursebook 1, we learnt that, in a histogram, when using probability density to scale the vertical axis, the total area of the columns is equal to 1.

Using increasingly narrow class intervals to draw a histogram of these data, and using probability density to find the heights of the bars, the graph increasingly approximates a curve and the area under the curve equals 1.

## Probability density functions

Consider a situation in which you are waiting for a taxi to arrive. The taxi is due to arrive within 8 minutes, but you do not know when during those 8 minutes it will arrive. The following three graphs represent three possible distributions of arrival times of the taxi.

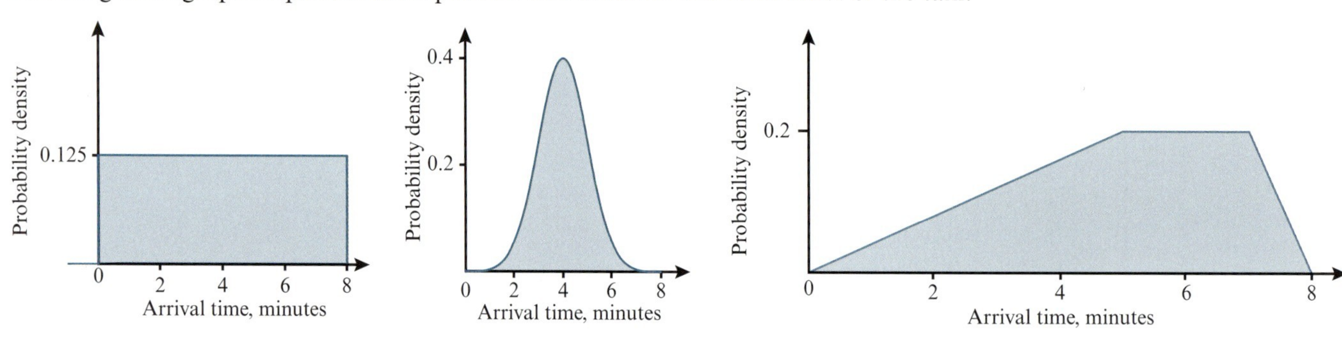


In the first graph, the arrival time of the taxi is equally likely at any point within the 8 minutes.

The second graph shows a normally distributed arrival time, so the taxi is most likely to arrive near the middle of the 8 minutes, where the graph peaks, and less likely at either end of the 8-minute time interval.



The third graph shows the most likely arrival time of the taxi as being between 5 minutes and 7 minutes, the period of time in which the probability is greatest.

Let's look in more detail at just one of these graphs, the final graph in the example 'taxi arrival time'.

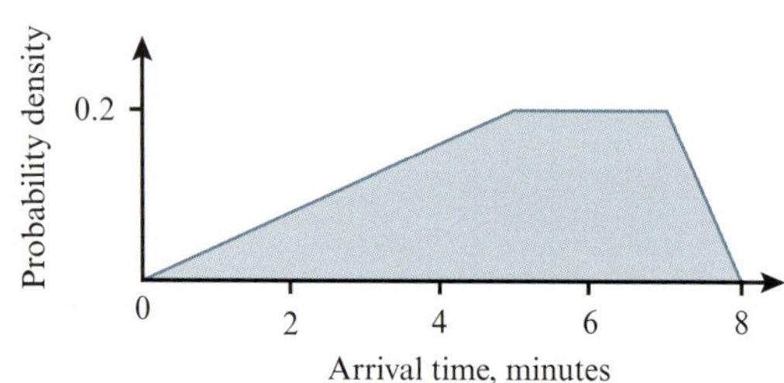


The area under the graph represents the total probability of the event occurring; that is, taxi arrival time. We can check that this is a PDF by calculating the area under the whole graph.

Using geometry: the shape of the graph is a trapezium; area $= \frac{1}{2} \times 0.2 \times (2 + 8) = 1$, as required.

Using calculus: we first need to know the function of the graph. The function for arrival time for this graph is defined as follows:

$$f(x) = \begin{cases} \dfrac{x}{25} & 0 \leqslant x < 5 \\[2ex] \dfrac{1}{5} & 5 \leqslant x < 7 \\[2ex] \dfrac{(8 - x)}{5} & 7 \leqslant x \leqslant 8 \\[2ex] 0 & \text{otherwise} \end{cases}$$

**REWIND**

Recall, from Chapter 8 in Probability & Statistics Coursebook 1, that a graph modelling the probability distribution of a continuous random variable is a **probability density function (PDF)**.

**REWIND**

You learnt about calculus in Pure Mathematics 1 Coursebook, and about geometry from IGCSE® / O Level.

Notice that since the graph is made up from three straight-line segments, its probability function is defined across three specified intervals.

Using calculus, we find the area under the graph by integrating the three functions in their corresponding intervals.

$$\text{Area} = \int_0^5 \frac{x}{25}\,\mathrm{d}x + \int_5^7 \frac{1}{5}\,\mathrm{d}x + \int_7^8 \frac{1}{5}(8-x)\,\mathrm{d}x = \left[\frac{x^2}{50}\right]_0^5 + \left[\frac{x}{5}\right]_5^7 + \left[\frac{1}{5}\left(8x - \frac{x^2}{2}\right)\right]_7^8$$

$$= \left(\frac{1}{2} - 0\right) + \left(\frac{7}{5} - 1\right) + \left(\frac{1}{5}\left(32 - 31\tfrac{1}{2}\right)\right) = \frac{1}{2} + \frac{2}{5} + \frac{1}{10} = 1, \text{ as required.}$$

**KEY POINT 4.1**

A graph, $\mathrm{f}(x)$, representing a continuous random variable is the probability density function (PDF).

The PDF has the following properties:

- It cannot be negative since you cannot have a negative probability; $\mathrm{f}(x) \geqslant 0$.
- Total probability of all outcomes = 1; hence, $\int_{-\infty}^{\infty} \mathrm{f}(x)\,\mathrm{d}x = 1$.

In many situations, the data are defined across a specified interval or across specified intervals, outside of which $\mathrm{f}(x) = 0$.

## Calculating probabilities

We can calculate the probability that the taxi arrival time will be in a given interval, but we cannot calculate the probability of an exact arrival time for the taxi; that is, we cannot calculate the probability of an exact outcome.

For example, consider the lifetime best performance for running the 10 000 metres for the athletes Tegla Loroupe from Kenya and Lornah Kipllagat from the Netherlands. Their lifetime best performances are both 30:37.26; that is, 30 minutes 37.26 seconds, but do they really have *exactly the same* lifetime best performance time? Performance time is given correct to the nearest hundredth of a second. The actual time may have been 30:37.264 or 30:37.259 or 30:37.256111 or ..... the possibilities are infinite. The performance time 30:37.26 is actually a range of times from 30:37.255 to 30:37.265. So, it follows that with a continuous random variable we can only calculate the probability of a range of values.



**KEY POINT 4.2**

- With continuous random variables, each individual value has zero probability of occurring.

  For a continuous random variable with PDF $\mathrm{f}(x)$, $\mathrm{P}(X = a) = 0$.

- Because we cannot find the probability of an exact value, when finding the probability in a given interval it does not matter whether you use $<$ or $\leqslant$.

  $\mathrm{P}(a < x < b) = \mathrm{P}(a \leqslant x < b) = \mathrm{P}(a < x \leqslant b) = \mathrm{P}(a \leqslant x \leqslant b)$

Note that this does not imply that $X$ cannot take the value $a$, it just means the probability of the exact value $a$ is zero.

To find the probability within a specified interval, we use calculus, more specifically integration, or sometimes geometry.

Let us consider again the final graph in the example 'taxi arrival time'. If we want to find the probability that the taxi arrives between 4 minutes and 6 minutes, then the shaded area in the following graph represents this probability.

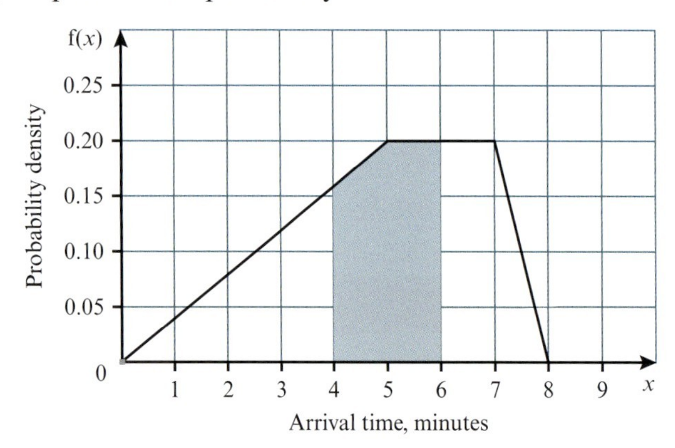


The shaded area consists of a trapezium and a rectangle. We can find the area, the probability of the taxi arriving between 4 and 6 minutes, using the formulae for area of a trapezium and for a rectangle.

$$P(4 < x < 6) = \frac{1}{2} \times 1 \times (0.16 + 0.2) + 0.2 \times 1 = 0.38$$

Alternatively, we can integrate $f(x)$ in the interval 4 to 6 to calculate the area under the graph.



$$f(x) = \begin{cases} \dfrac{x}{25} & 0 \leqslant x < 5 \\ \dfrac{1}{5} & 5 \leqslant x < 7 \\ \dfrac{(8-x)}{5} & 7 \leqslant x \leqslant 8 \\ 0 & \text{otherwise} \end{cases}$$

$$P(4 < x < 6) = \int_4^5 \frac{x}{25}\,dx + \int_5^6 \frac{1}{5}\,dx = \left[\frac{x^2}{50}\right]_4^5 + \left[\frac{x}{5}\right]_5^6 = \left(\frac{25}{50} - \frac{16}{50}\right) + \left(\frac{6}{5} - 1\right) = 0.38$$

**TIP**

You only need to integrate the parts of the function within the interval of interest.

**KEY POINT 4.3**

The probability of $X$ lying in the interval $(a, b)$ is given by the area under the graph between $a$ and $b$. That is:

$$P(a < X < b) = \int_a^b f(x)\,dx$$

**WORKED EXAMPLE 4.1**

The height reached by water erupting from a broken water pipe, $X$ metres, is modelled by the following PDF.

$$f(x) = \begin{cases} \dfrac{x^2}{k} & 1 \leqslant x \leqslant 10 \\ 0 & \text{otherwise} \end{cases}$$

**a** Show that $k = 333$.

**b** Sketch the graph of $f(x)$.

**c** Find the probability that the water reaches a height of at least 6 m.

**Answer**

**a** $\int_1^{10} \frac{x^2}{k}\, dx = 1$

For this to be a PDF, the area under the curve = 1.

$$\left[\frac{x^3}{3k}\right]_1^{10} = 1$$

$$\frac{1000}{3k} - \frac{1}{3k} = 1$$

$$\frac{999}{3k} = 1$$

$$k = 333$$

**b**

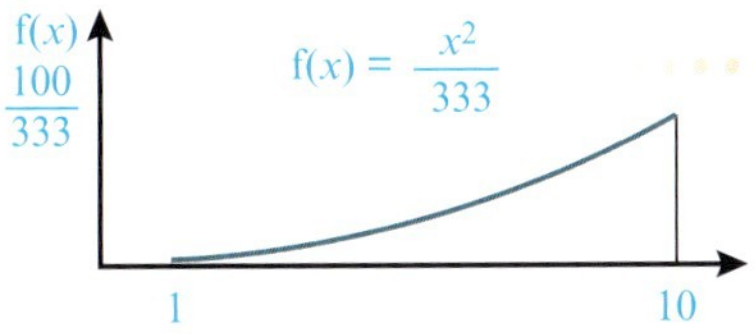


You only need to show the part of the graph between the limits. Show relevant values on both axes.

For $x = 10$, $f(x) = \frac{10^2}{333} = \frac{100}{333}$

**c** $P(X > 6) = \int_6^{10} \frac{x^2}{333}\, dx$

To find the probability, calculate the area of the graph from $x = 6$ to $x = 10$. We can use the result of our integration in part **a** to save repeated working.

$$= \left[\frac{x^3}{999}\right]_6^{10}$$

$$= \frac{1000}{999} - \frac{216}{999}$$

$$= \frac{784}{999}$$



## WORKED EXAMPLE 4.2

A brand of laptop battery has a lifetime of $X$ years. It is suggested that the variable $X$ can be modelled by the following PDF.

$$f(x) = \begin{cases} \frac{2}{x^2} & x \geqslant 2 \\ 0 & \text{otherwise} \end{cases}$$

**a** Sketch the graph of $f(x)$.

**b** Show that $f(x)$ has the properties required of a probability density function.

**c** Find the probability that the battery lasts for more than 3 years.

**d** Work out the value $m$ such that $P(X < m) = \frac{1}{2}$.

**Answer**

**a**

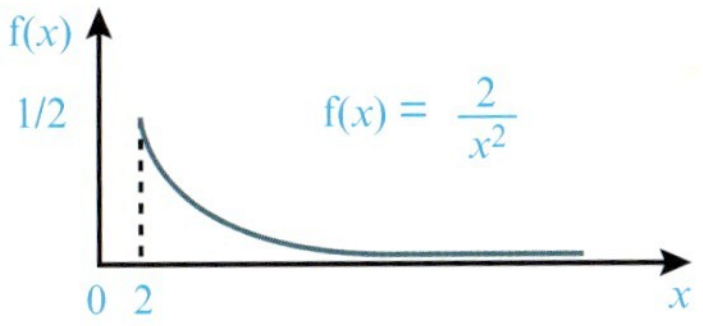


We only need to show the part of the graph between the limits. Show relevant values on both axes.

For $x = 10$, $f(x) = \frac{2}{2^2} = \frac{1}{2}$

**b** Area $= \int_2^{\infty} \frac{2}{x^2}\,dx = \left[\frac{-2}{x}\right]_2^{\infty}$

$= \frac{-2}{\infty} - \frac{-2}{2}$

$= 0 - (-1) = 1$

The PDF can never be negative as $x^2$ is always positive. We also need to show the area under the graph = 1.

**c** $P(X > 3) = \int_3^{\infty} \frac{2}{x^2}\,dx = \left[\frac{-2}{x}\right]_3^{\infty}$

$= \frac{-2}{\infty} - \frac{-2}{3}$

$= 0 - \frac{-2}{3} = \frac{2}{3}$

**d** $P(X < m) = \int_2^{m} \frac{2}{x^2}\,dx = \left[\frac{-2}{x}\right]_2^{m}$

$= \frac{-2}{m} - \frac{-2}{2}$

$= \frac{m-2}{m} = \frac{1}{2}$

$m = 4$

Use $m$ as the upper limit of the integral.



## EXERCISE 4A

**1** The continuous random variable $X$ has the following probability density function:

$$f(x) = \begin{cases} kx^4 + \frac{1}{5} & 0 < x < 2 \\ 0 & \text{otherwise} \end{cases}$$

**a** Show that $k = \frac{3}{32}$.

**b** Sketch the graph of $f(x)$.

**c** Find $P(0 < X < 1)$.

**2 a** Given that $f(x) = \begin{cases} \frac{x^3}{60} & 2 < x < 4 \\ 0 & \text{otherwise} \end{cases}$, show that $f(x)$ has the properties of a probability density function and sketch $f(x)$.

**b** Find $P(X > 3)$.

**c** Find $P(X < 2)$.

**3** The continuous random variable $X$ has the following probability density function:

$$f(x) = \begin{cases} \frac{3}{32}x^2 & 0 \leqslant x < 2 \\ \frac{1}{24}(13 - 2x) & 2 \leqslant x \leqslant 5 \\ 0 & \text{otherwise} \end{cases}$$

**a** Show that $f(x)$ has the properties of a probability density function, and sketch $f(x)$.

**b** Find: **i** $P(X > 4)$ **ii** $P(1 < X < 3)$

**4** $f(x) = \begin{cases} \dfrac{2(x-1)}{3} & 1 \leqslant x \leqslant 2 \\ \dfrac{4-x}{3} & 2 < x \leqslant 4 \\ 0 & \text{otherwise} \end{cases}$

**a** Show that $f(x)$ has the properties of a probability density function and sketch $f(x)$.

**b** Find $P(X < 3)$.

**c** Find the value of $k$ such that $P(X < k) = \dfrac{1}{3}$.

**5** $f(x) = \begin{cases} \dfrac{2x}{3} & 0 \leqslant x \leqslant 1 \\ \dfrac{2}{3} & 1 < x \leqslant 2 \\ 0 & \text{otherwise} \end{cases}$

Find the value of $m$ such that $P(X < m) = 0.5$.

**PS** **6** Jenna and Alex share the use of a car. Jenna uses the car 75% of the time and Alex 25% of the time. After Jenna uses the car, the number of litres of fuel left in the tank is given by the continuous random variable $X$, given by:

$$f(x) = \begin{cases} \dfrac{1}{50} & 15 \leqslant x \leqslant 65 \\ 0 & \text{otherwise} \end{cases}$$

After Alex uses the car, the number of litres of fuel left in the tank is given by the continuous random variable $Y$, given by:

$$f(y) = \begin{cases} \dfrac{1}{25} & 0 \leqslant y \leqslant 25 \\ 0 & \text{otherwise} \end{cases}$$

Jenna, not knowing who used the car last, checks the amount of fuel left in the tank. Find the probability there is less than 20 litres of fuel left in the tank.

**PS** **7** Mia breaks glasses at the rate of four per week. Let $T$ be the time, in weeks, between successive breakages of glasses. Then:

$$f(t) = \begin{cases} 4\,e^{-4t} & t \geqslant 0 \\ 0 & \text{otherwise} \end{cases}$$

What is the probability that a week goes by without Mia breaking any glasses?

## 4.2 Finding the median and other percentiles of a continuous random variable

In Worked example 4.2 part **d**, you found the value within the interval such that the probability $= \frac{1}{2}$. The question required finding the value $m$ such that $P(X < m) = \frac{1}{2}$. Since the total probability = 1, not only is $P(X < m) = \frac{1}{2}$, but also $P(X > m) = \frac{1}{2}$. The value $m$ must therefore be the median, the value of $X$ where the probability is exactly half.

**KEY POINT 4.4**

The median, $m$, of a continuous random variable is that value for which

$$P(X < m) = \int_{-\infty}^{m} f(x)\ dx = \frac{1}{2}$$

**REWIND**

Recall from Probability & Statistics Coursebook 1 that the median is the middle value for a set of data. For a PDF, the total probability = 1, so the middle value or median occurs when the area under the graph up to that point $= \frac{1}{2}$.

Another way to express the median is to refer to it as the 50th percentile. We can find other percentiles of continuous random variables using a similar method to that given in Key point 4.4, with the fraction $\frac{1}{2}$ replaced by the relevant percentage.

For example, if the value $r$ of a continuous random variable is that value that represents the 30th percentile, then $\int_{-\infty}^{r} f(x)\ dx = 30\%$ or 0.3.

**WORKED EXAMPLE 4.3**

For any given time I arrive at a bus stop, the arrival time of a bus can be modelled by the continuous random variable $X$, whose PDF f($x$) is given by:

$$f(x) = \begin{cases} \dfrac{x^2}{72} & 0 \leqslant x \leqslant 6 \\ 0 & \text{otherwise} \end{cases}$$

Find the median and interquartile range (IQR) of the arrival time of the bus.



**Answer**

Let $m$ be the median, $\int_0^m \frac{x^2}{72}\ dx = 0.5$

The median, 50th percentile, is such that the area up to the median value, $m$, is 50% or 0.5.

An alternative calculation for the upper quartile is to look at the last 25% of the area, and you would calculate $\int_v^6 \frac{x^2}{72}\ dx = 0.25$.

$$\left[\frac{x^3}{216}\right]_0^m = 0.5$$

$$\frac{m^3}{216} = 0.5$$

$$m^3 = 108$$

$$m = 4.76 \text{ (2 decimal places)}$$

Let $t$ be the lower quartile. Then:

$$\int_0^t \frac{x^2}{72}\ dx = 0.25$$

$$\left[\frac{x^3}{216}\right]_0^t = 0.25$$

$$\frac{t^3}{216} = 0.25$$

$$t^3 = 54$$

$$t = 3.78 \text{ (2 decimal places)}$$

Let $v$ be the upper quartile. Then:

$$\int_0^v \frac{x^2}{72}\,dx = 0.75$$

$$\left[\frac{x^3}{216}\right]_0^v = 0.75$$

$$\frac{v^3}{216} = 0.75$$

$$v^3 = 162$$

$$v = 5.45 \text{ (2 decimal places)}$$

$\text{IQR} = \text{UQ} - \text{LQ} = 5.45 - 3.78 = 1.67$

**WORKED EXAMPLE 4.4**

A continuous random variable, $X$, has PDF $f(x)$ given by:

$$f(x) = \begin{cases} \frac{2}{21}(6 - x) & 1 \leqslant x \leqslant 4 \\ 0 & \text{otherwise} \end{cases}$$

**a** Show that the median value, $m$, is given by $2m^2 - 24m + 43 = 0$.

**b** Find the value of $m$, correct to 2 decimal places.

**Answer**

**a**

$$\int_1^m \frac{2}{21}(6 - x)\,dx = 0.5$$

Area under graph up to median value = 0.5.

$$\left[\frac{2}{21}\left(6x - \frac{x^2}{2}\right)\right]_1^m = 0.5$$

$$\frac{2}{21}\left(\left(6m - \frac{m^2}{2}\right) - \left(6 - \frac{1}{2}\right)\right) = 0.5$$

$$12m - m^2 - 11 = 10.5$$

$2m^2 - 24m + 43 = 0$, as required.

You must show sufficient working when the result is given in the question.

An alternative calculation, using the area beyond/to the right of the median, is:

$$\int_m^4 \frac{2}{21}(6 - x)\,dx = 0.5$$

**EXPLORE 4.2**

Show that you get the same result with this integral.

**b**

$$2m^2 - 24m + 43 = 0$$

Use the quadratic formula or complete the square.

$$m^2 - 12m + 21.5 = 0$$

$$(m - 6)^2 - 36 + 21.5 = 0$$

$$(m - 6)^2 = 14.5$$

$$m = 6 \pm \sqrt{14.5}$$

$$m = 2.19$$

Choose the solution that falls within the domain of the PDF.

**WORKED EXAMPLE 4.5**

A continuous random variable, $X$, has the following PDF:

$$f(x) = \begin{cases} \frac{1}{27}x^2 & 0 \leqslant x \leqslant 3 \\ \frac{1}{3}(5 - x) & 3 < x \leqslant 5 \\ 0 & \text{otherwise} \end{cases}$$

**a** Find the median.

**b** Find the lower quartile.

**Answer**

**a** $\int_0^3 \frac{x^2}{27} \, dx = \left[\frac{x^3}{81}\right]_0^3 = \frac{1}{3}$

The PDF is defined across more than one interval. We need to first work out in which interval the median lies.

$\frac{1}{3} < \frac{1}{2}$; hence, the median lies in the second part.

$$\int_m^5 \frac{1}{3}(5 - x) \, dx = 0.5$$

$$\left[\frac{1}{3}\left(5x - \frac{x^2}{2}\right)\right]_m^5 = 0.5$$

$$\left(25 - \frac{25}{2}\right) - \left(5m - \frac{m^2}{2}\right) = 1.5$$

$$m^2 - 10m + 22 = 0$$

$$(m - 5)^2 - 25 + 22 = 0$$

$$m = 5 \pm \sqrt{3}$$

$$m = 3.27$$

The alternative integral is:

$\frac{1}{3} + \int_3^m \frac{1}{3}(5 - x) \, dx = 0.5$

If you use this, remember to include $\frac{1}{3}$ from the first part of the PDF.

Form and solve a quadratic equation, choosing the value within the relevant range.

Sometimes a sketch may help us to decide where the relevant value will be.

**b** Let $t$ be the lower quartile. Then:

$$\left[\frac{x^3}{81}\right]_0^t = 0.25$$

$$\frac{t^3}{81} = 0.25$$

$$t = 2.73$$

The lower quartile, 25th percentile, 0.25 or $\frac{1}{4} < \frac{1}{3}$; hence, the value will lie in the first part of the PDF.

Use the integral you've already calculated in part **a**.



**EXERCISE 4B**

1 A continuous random variable, $X$, has the following probability density function:

$$f(x) = \begin{cases} k(x - 2) & 2 \leqslant x < 5 \\ k(7 - x) & 5 \leqslant x \leqslant 7 \\ 0 & \text{otherwise} \end{cases}$$

**a** Show that $k = \frac{2}{13}$.

**b** Find $P(2 < X < 4)$.

c Find $P(3 < X < 6)$.

d Find the median value, giving your answer in surd form.

e Find the value of $x$ such that $P(X < x) = 20\%$, giving your answer in surd form.

2 A continuous random variable, $X$, has the following probability density function:

$$f(x) = \begin{cases} \frac{2}{25}(5 - x) & 0 \leqslant x \leqslant 5 \\ 0 & \text{otherwise} \end{cases}$$

a Find $P(X \leqslant 2)$.

b Show that the median is $5 - \sqrt{\frac{(50)}{2}}$.

c Calculate the interquartile range.

M 3 Two sisters make contact using an internet messaging service. The length of time for which they are logged on, in minutes, is modelled by the random variable $T$ with probability density function:

$$f(t) = \begin{cases} \frac{1}{k}(40 - t) & 10 \leqslant t \leqslant 30 \\ 0 & \text{otherwise} \end{cases}$$

where $k$ is a constant.

a Verify that $k = 400$.

b Find the probability that the time that they are logged on for is less than 15 minutes.

c Calculate the median value for time they are logged on for.

d Give a reason why this model may not be realistic.



PS 4 Patients arriving at a dental surgery wait in reception before seeing the dentist. The duration of their waiting time, $T$ minutes, is modelled using the following probability density function:

$$f(t) = \begin{cases} kt & 0 \leqslant t \leqslant 22 \\ 0 & \text{otherwise} \end{cases}$$

where $k$ is a constant.

a Sketch the graph of $f(t)$ and show that $k = \frac{1}{242}$.

b Find the probability that a randomly chosen patient waits in reception for longer than 5 minutes.

c The dental surgery claims that 80% of patients wait for fewer than 20 minutes. Show that the claim is justified.

M 5 A continuous variable, $X$, has the following probability density function:

$$f(x) = \begin{cases} kx & 0 \leqslant x < 2.5 \\ k(5 - x) & 2.5 \leqslant x \leqslant 5 \\ 0 & \text{otherwise} \end{cases}$$

a Verify that $k = \frac{4}{25}$.

b Find $P(2 < X < 4)$.

**c** Sketch $f(x)$ and use your graph to:

**i** explain why the modal value of $X$ is 2.5

**ii** state the median value of $X$.

**6** The time to failure of a particular type of toaster, in years, can be modelled by the continuous random variable $X$, which has the following probability density function:

$$f(x) = \begin{cases} \frac{1}{6}(1+4x) & 0 \leqslant x < 1 \\ \frac{5}{6\sqrt{x}} & 1 \leqslant x \leqslant 1.69 \\ 0 & \text{otherwise} \end{cases}$$

**a** Find the probability that a toaster of this type fails between 3 months and 18 months.

**b** A store sells 1600 of this type of toaster with a 1-year warranty. What is the expected number of toasters that will fail during the warranty period?

## 4.3 Finding the expectation and variance

Consider Worked example 4.3. Suppose there is no bus timetable and the only information you have about the buses along this route is that the arrival time of a bus can be modelled by the continuous random variable $X$, whose PDF $f(x)$ is given by:



$$f(x) = \begin{cases} \frac{x^2}{72} & 0 \leqslant x \leqslant 6 \\ 0 & \text{otherwise} \end{cases}$$

We want to find out the expected, mean, waiting time and the variance in waiting times for anyone arriving at the bus stop and using this bus service.

With discrete random variables to find the mean, we calculate $E(X) = \sum_{\text{all } x} xP(x)$.

However, with continuous random variables, $P(x) = 0$ for all specific values of $X$, so it is not possible to use this formula as it is given.

This is the graph of the PDF $f(x) = \frac{x^2}{72}$.

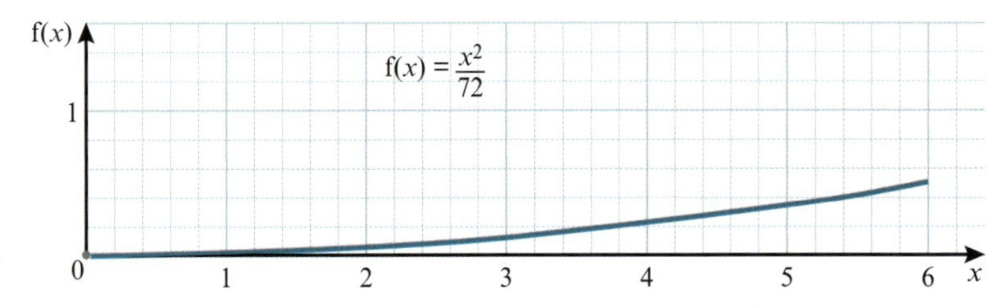


If we find the probability of a very small interval, $\delta x$, on the graph, this would be a very small strip, approximately a rectangle with width $\delta x$ and height $f(x_0)$.

From Key point 4.3 we can find the probability represented by this small strip:

$P(x_0 < x < x_0 + \delta x) \approx f(x_0)\delta x$

Hence, using the formula for the mean of $x$:

$$E(X) = \sum_{\text{all } x} xP(x) \approx \sum_{\text{all } x} xf(x)\delta(x) = \int_{-\infty}^{\infty} x\, f(x)\, dx$$

The same reasoning works to find the variance.

$$E(X^2) \approx \sum_{\text{all } x} x^2 \, P(x) = \int_{-\infty}^{\infty} x^2 \, f(x) \, dx$$

$$\text{Var}(X) = E(X^2) - \{E(X)\}^2 = \int_{-\infty}^{\infty} x^2 \, f(x) \, dx - \left\{\int_{-\infty}^{\infty} x \, f(x) \, dx\right\}^2$$

Returning to Worked example 4.3, we can now find the mean and variance for the bus arrival time for the continuous random variable $X$, whose PDF $f(x)$ is given by:

$$f(x) = \begin{cases} \dfrac{x^2}{72} & 0 \leqslant x \leqslant 6 \\ 0 & \text{otherwise} \end{cases}$$

$$x \, f(x) = x \frac{x^2}{72} = \frac{x^3}{72}$$

$$\text{Mean, } E(X) = \int_0^6 \frac{x^3}{72} \, dx = \left[\frac{x^4}{288}\right]_0^6 = \left(\frac{1296}{288}\right) - (0) = 4.5$$

To find the variance, first work out $E(X^2)$:

$$x^2 \, f(x) = x^2 \frac{x^2}{72} = \frac{x^4}{72}$$

$$E(X^2) = \int_0^6 \frac{x^4}{72} \, dx = \left[\frac{x^5}{360}\right]_0^6 = \left(\frac{7776}{360}\right) = 21.6$$

$$\text{Var}(X) = E(X^2) - \{E(X)\}^2 = 21.6 - 4.5^2 = 21.6 - 20.25 = 1.35$$



**KEY POINT 4.5**

For continuous random variables with PDF $f(x)$:

$$E(X) = \int_{-\infty}^{\infty} x \, f(x) \, dx \text{ and } \text{Var}(X) = \int_{-\infty}^{\infty} x^2 \, f(x) \, dx - \left\{\int_{-\infty}^{\infty} x \, f(x) \, dx\right\}^2$$

**WORKED EXAMPLE 4.6**

The time, in minutes, taken by students to answer a question in a multiple-choice test has a PDF $f(t)$ as follows:

$$f(t) = \begin{cases} \dfrac{1}{99}(40 - t^2) & 1 \leqslant t \leqslant 4 \\ 0 & \text{otherwise} \end{cases}$$

Work out the mean and variance of the time taken.

**Answer**

$$E(T) = \int_1^4 \frac{1}{99}(40t - t^3) \, dt = \frac{1}{99}\left[20t^2 - \frac{t^4}{4}\right]_1^4$$

$$= \frac{1}{99}\left(\left(20 \times 4^2 - \frac{4^4}{4}\right) - \left(20 - \frac{1}{4}\right)\right) = 2\tfrac{17}{44}$$

To carry out the integration, we need to multiply each term in $f(t)$ by $t$ and by $t^2$ to find $E(T)$ and $E(T^2)$, respectively.

If we need to also find the variance, give the answer to the mean as an exact fraction rather than a decimal or work to at least four significant figures.

$$E(T^2) = \int_1^4 \frac{1}{99}(40t^2 - t^4)\ dt = \left[\frac{1}{99}\left(\frac{40}{3}t^3 - \frac{t^5}{5}\right)\right]_1^4$$

To find the variance, first calculate $E(T^2)$.

$$= \frac{1}{99}\left(\left(\frac{40}{3} \times 4^3 - \frac{4^5}{5}\right) - \left(\frac{40}{3} - \frac{1}{5}\right)\right) = 6\,\tfrac{23}{55}$$

$$\text{Var}(T) = 6\,\tfrac{23}{55} - \left(2\,\tfrac{17}{44}\right)^2 = 0.723$$

## WORKED EXAMPLE 4.7

The arrival time, $x$, of a taxi is a continuous random variable $X$, whose PDF $f(x)$ is given by:

$$f(x) = \begin{cases} \dfrac{x}{25} & 0 \leqslant x < 5 \\ \dfrac{1}{5} & 5 \leqslant x \leqslant 7 \\ \dfrac{(8 - x)}{5} & 7 < x \leqslant 8 \\ 0 & \text{otherwise} \end{cases}$$

The taxi firm states that, on average, a taxi will arrive within 5 minutes. Is this statement true or false?

**Answer**

$$\int_0^5 \frac{x^2}{25}\ dx + \int_5^7 \frac{1}{5}x\ dx + \int_7^8 \frac{1}{5}(8x - x^2)\,dx$$

Taking the average to be the mean average, we need to calculate $E(X)$.

The PDF is defined in parts, so we need to multiply each part by $x$ and then integrate each part within its limits.

$$= \left[\frac{x^3}{75}\right]_0^5 + \left[\frac{1}{10}x^2\right]_5^7 + \left[\frac{1}{5}\left(4x^2 - \frac{x^3}{3}\right)\right]_7^8$$

$$= \left(\frac{125}{75} - 0\right) + \left(\frac{49}{10} - \frac{25}{10}\right) +$$

$$\frac{1}{5}\left(256 - 196 - \left(\frac{512}{3} - \frac{343}{3}\right)\right) = 4.8$$

$4.8 < 5$ so the statement is true.

## EXERCISE 4C

1 A continuous random variable, $X$, has probability density function as follows:

$$f(x) = \begin{cases} \dfrac{6}{125}(5x - x^2) & 0 \leqslant x \leqslant 5 \\ 0 & \text{otherwise} \end{cases}$$

Find the mean and variance of $X$.

2 A continuous random variable $X$ has the following probability density function:

$$f(x) = \begin{cases} \dfrac{3}{32}x^2 & 0 \leqslant x < 2 \\ \dfrac{1}{24}(13 - 2x) & 2 \leqslant x \leqslant 5 \\ 0 & \text{otherwise} \end{cases}$$

Find the mean and variance of $X$.

3 A continuous random variable, $X$, has probability density function given by:

$$f(x) = \begin{cases} kx^2 & 0 \leqslant x < 2 \\ k(6-x) & 2 \leqslant x \leqslant 6 \\ 0 & \text{otherwise} \end{cases}$$

where $k$ is a constant.

a Show that $k = \frac{3}{32}$.

b Calculate the mean and variance of $X$.

PS 4 A forester packs kindling in bags to sell. The mass of kindling in a bag, $X$ kg, is a continuous random variable with probability density function given by:

$$f(x) = \begin{cases} k(3x^2 - 4x + 2) & 3 \leqslant x \leqslant 5 \\ 0 & \text{otherwise} \end{cases}$$

a Show that $k = \frac{1}{70}$.

b Samuel buys a bag of kindling. What is the expected mass of kindling Samuel will get in the bag?

c The forester has 100 bags of kindling to sell. He charges \$3 for each bag of kindling with a mass of 4 kg or more, otherwise he charges \$2 per bag. Calculate the amount the forester may expect to receive for these 100 bags of kindling.

PS 5 An environmentalist develops an index for the quality of pond water. The quality index of a sample of pond water is a continuous random variable $X$, having the following probability density function:

$$f(x) = \begin{cases} 12(x - 2x^2 + x^3) & 0 \leqslant x \leqslant 1 \\ 0 & \text{otherwise} \end{cases}$$

a Find the expected value of the quality index.

b Frogspawn survive best in ponds where the quality index of the pond water is greater than 0.8. If the environmentalist checks a random sample of 150 ponds, in how many ponds will frogspawn survive best?

PS 6 A machine produces components for industrial use. The machine has an inbuilt device to accept components of lengths within certain limits. The acceptable lengths of components produced may be considered a continuous random variable $X$, with the following probability density function:

$$f(x) = \begin{cases} \frac{1}{20}(10x - x^3) & 1 \leqslant x \leqslant 3 \\ 0 & \text{otherwise} \end{cases}$$

a Find the mean and standard deviation of $X$.

b Find the probability that the length of a component is less than the mean.

c A customer requires components with lengths that are within one standard deviation of the mean. In a batch of 1000 components, how many will be acceptable to this customer? Give your answer to an appropriate degree of accuracy.

M 7 Let $T$ be the lifetime, in years, of a particular type of battery. Then:

$$f(t) = \begin{cases} 0.2\,e^{-0.2t} & t \geqslant 0 \\ 0 & \text{otherwise} \end{cases}$$

a Find the probability that the battery lasts longer than 1.5 years.

b Find the mean and variance of the number of years the battery will last.

**EXPLORE 4.3**

Consider again the first graph for the taxi arrival time in the subsection on probability distribution functions in Section 4.1. This graph shows that the arrival time is equally likely at any point during the 8 minutes. The shape of the graph gives a special type of continuous random variable, which is said to have a rectangular (or uniform) distribution, as shown in the following diagram.

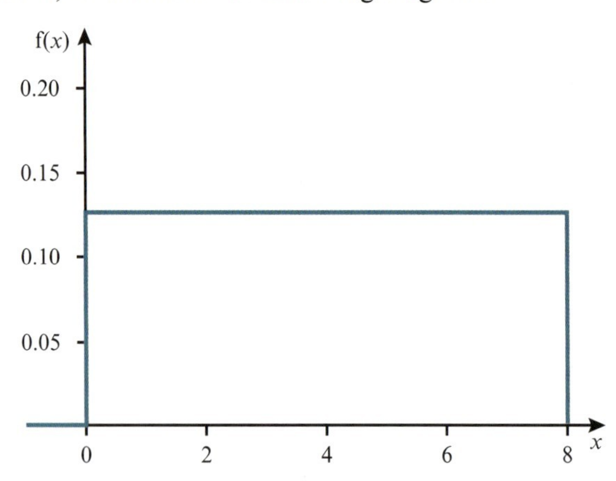


The PDF of this graph is:

$$f(x) = \begin{cases} \frac{1}{8} & 0 \leqslant x \leqslant 8 \\ 0 & \text{otherwise} \end{cases}$$



**a** Without doing any calculations, explain why:

**i** the mean $E(X) = 4$

**ii** the median $= 4$

**iii** there is no mode.

**b** In the general case, the PDF $f(x)$ of a continuous random variable $X$ on the interval from $a$ to $b$ that has a rectangular distribution, is given by:

$$f(x) = \begin{cases} \frac{1}{b-a} & a \leqslant x \leqslant b \\ 0 & \text{otherwise} \end{cases}$$

**i** Draw the graph of this distribution.

**ii** In terms of $a$ and $b$, state the mean, median and mode, and work out a formula for the variance.

**EXPLORE 4.4**

A continuous random variable, $T$, is given by the following PDF $\mathrm{f}(t)$:

$$\mathrm{f}(t) = \begin{cases} \lambda e^{-\lambda t} & t \geqslant 0 \\ 0 & \text{otherwise} \end{cases}$$

where $\lambda$ is a positive constant and is said to follow an exponential distribution.

**a** Sketch the graph of $\mathrm{f}(t)$ and show that $T$ is a random variable.

**b** Find $\mathrm{P}(T < a)$ and $\mathrm{P}(T > a)$.

**c** Show that $\mathrm{E}(T) = \dfrac{1}{\lambda}$ and $\mathrm{Var}(T) = \dfrac{1}{\lambda^2}$.

An example of a continuous random variable that follows an exponential distribution is the 'wait times' between successive events in a Poisson distribution. Exercise 4A question 7 and Exercise 4C question 7 are exponential distributions. Use the general results you have just found to check your solutions to these questions.

## Checklist of learning and understanding



If $X$ is a continuous random variable with probability density function (PDF) $\mathrm{f}(x)$:

- $\int_{-\infty}^{\infty} \mathrm{f}(x)\,\mathrm{d}x = 1$
- $\mathrm{f}(x) \geqslant 0$ for all $x$
- $\mathrm{P}(a < X < b) = \int_a^b \mathrm{f}(x)\,\mathrm{d}x$
- $\mathrm{P}(X = a) = 0$
- $\mathrm{E}(x) = \int_{\text{all } x} x\,\mathrm{f}(x)\,\mathrm{d}x$
- $\mathrm{Var}(x) = \int_{\text{all } x} x^2\,\mathrm{f}(x)\,\mathrm{d}x - \left\{\int_{\text{all } x} x\,\mathrm{f}(x)\,\mathrm{d}x\right\}^2$

## END-OF-CHAPTER REVIEW EXERCISE 4

PS **1** Rahini catches a bus to work from the same bus stop each morning. The arrival time, in minutes, of the bus Rahini catches each morning is modelled by the continuous random variable $T$ (which is measured from the moment that Rahini arrives at the bus stop) having the following probability density function:

$$f(t) = \begin{cases} \frac{1}{144}t^2 & 0 \leqslant t < 6 \\ \frac{1}{16}(10 - t) & 6 \leqslant t \leqslant 10 \\ 0 & \text{otherwise} \end{cases}$$

**a** Sketch the graph of $f(t)$. **[3]**

**b** Find the probability the bus arrives within 7 minutes. **[3]**

**c** What is the expected time, in minutes and seconds, that Rahini will wait at the bus stop? **[4]**

**2** A continuous random variable, $X$, has probability density function given by:

$$f(x) = \begin{cases} k(x - 3) & 3 \leqslant x < 7 \\ k(11 - x) & 7 \leqslant x \leqslant 11 \\ 0 & \text{otherwise} \end{cases}$$

**a** Show that $k = \frac{1}{16}$. **[3]**

**b** Sketch $f(x)$ and use your sketch to show that the median is 7. **[4]**



**c** Find the value of $a$ if $P(3 \leqslant X \leqslant a) = 0.75$. **[3]**

**d** Given that the mean is 7, calculate the variance of $X$. **[4]**

M **3** The arrival time, in minutes, of customers at a bank during a 90-minute interval, measured from the start of the time interval, is a continuous random variable $T$ with probability density function as follows:

$$f(t) = \begin{cases} \frac{1}{90} & 0 \leqslant t \leqslant 90 \\ 0 & \text{otherwise} \end{cases}$$

**a** Find the probability that a customer arrives during the first 30 minutes. **[3]**

**b** Sketch the graph of $f(t)$ and use your graph to explain why the probability of a customer arriving during any 30-minute interval is equal to your answer to part **a**. **[3]**

**c** State the expected arrival time. **[1]**

The time in minutes for a customer to be served at the bank is a continuous random variable $X$ with probability density function given by:

$$f(x) = \begin{cases} 0.5\,e^{-0.5x} & x > 0 \\ 0 & \text{otherwise} \end{cases}$$

**d** What is the probability that a customer requires between 2 minutes and 5 minutes to be served? **[3]**

**e** What is the probability that a customer arrives in the first 30 minutes and takes between 2 minutes and 5 minutes to be served? **[1]**

 4 The lifespan, in weeks, of a species of moth is a continuous random variable $T$ with probability density function given by:

$$f(t) = \begin{cases} \frac{1}{18}(9 - t^2) & 0 \leqslant t \leqslant 3 \\ 0 & \text{otherwise} \end{cases}$$

a Find the probability that a randomly chosen moth of this species lives for longer than 1 week. [3]

b Find the mean and variance of the lifespan of this species of moth. [7]

c Show that just over 66% of moths live for 10 days or fewer. [3]

 5

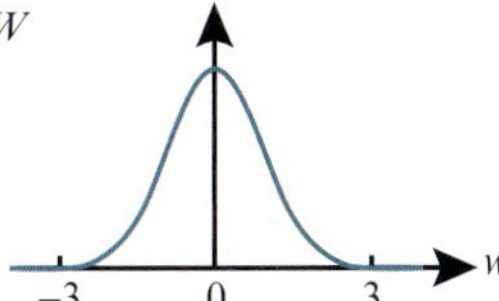

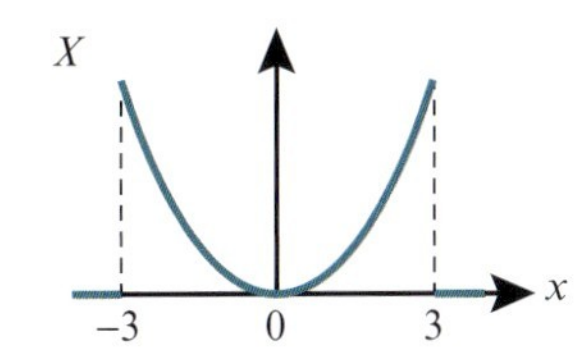

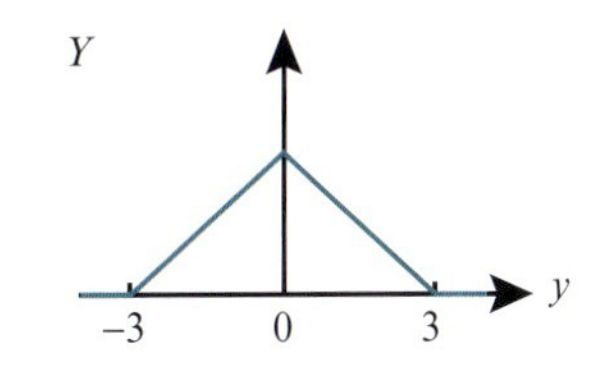

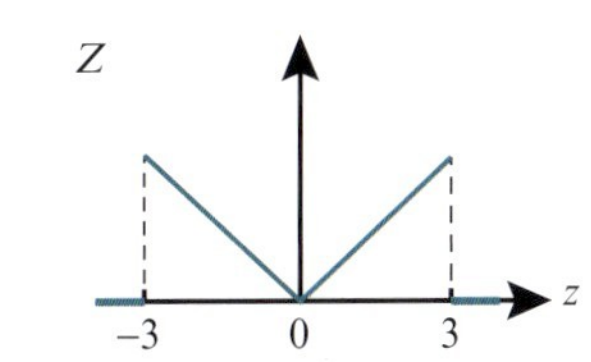


The diagrams show the probability density functions of four random variables $W$, $X$, $Y$ and $Z$. Each of the four variables takes values between –3 and 3 only, and their standard deviations are $\sigma_W$, $\sigma_X$, $\sigma_Y$ and $\sigma_Z$, respectively.

i List $\sigma_W$, $\sigma_X$, $\sigma_Y$ and $\sigma_Z$ in order of size, starting with the largest. [2]

ii The probability density function of $X$ is given by:

$$f(x) = \begin{cases} \frac{1}{18}x^2 & -3 \leqslant x \leqslant 3 \\ 0 & \text{otherwise} \end{cases}$$

a Show that $\sigma_X = 2.32$, correct to 3 significant figures. [3]

b Calculate $P(X > \sigma_X)$. [3]

c Write down the value of $P(X > 2\sigma_X)$. [1]

*Cambridge International AS & A Level Mathematics 9709 Paper 73 Q8 November 2016*

   6 a

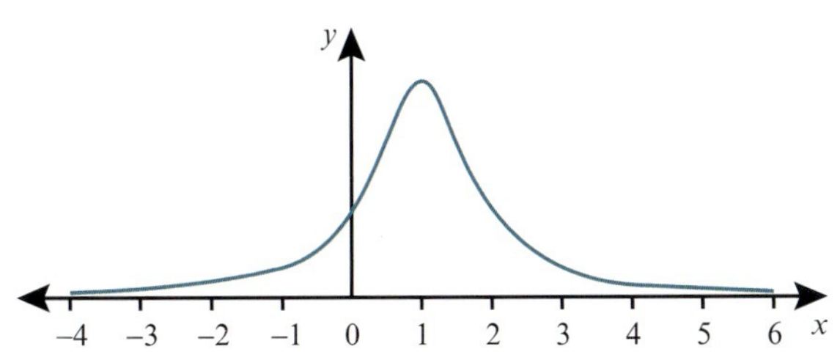


The diagram shows the graph of the probability density function of a variable $X$. Given that the graph is symmetrical about the line $x = 1$ and that $P(0 < X < 2) = 0.6$, find $P(X > 0)$. [2]

b A flower seller wishes to model the length of time that tulips last when placed in a jug of water. She proposes a model using the random variable $X$ (in hundreds of hours) with probability density function given by:

$$f(x) = \begin{cases} k(2.25 - x^2) & 0 \leqslant x \leqslant 1.5 \\ 0 & \text{otherwise} \end{cases}$$

where $k$ is a constant.

i Show that $k = \frac{4}{9}$. **[3]**

ii Use this model to find the mean number of hours that a tulip lasts in a jug of water. **[4]**

The flower seller wishes to create a similar model for daffodils. She places a large number of daffodils in jugs of water and the longest time that any daffodil lasts is found to be 290 hours.

iii Give a reason why $f(x)$ would not be a suitable model for daffodils. **[1]**

iv The flower seller considers a model for daffodils of the form:

$$g(x) = \begin{cases} c(a^2 - x^2) & 0 \leqslant x \leqslant a \\ 0 & \text{otherwise} \end{cases}$$

where $a$ and $c$ are constants. State a suitable value for $a$. (There is no need to evaluate $c$.) **[1]**

*Cambridge International AS & A Level Mathematics 9709 Paper 72 Q7 June 2016*

7 a The time for which Lucy has to wait at a certain traffic light each day is $T$ minutes, where $T$ has probability density function given by:

$$f(t) = \begin{cases} \frac{3}{2}t - \frac{3}{4}t^2 & 0 \leqslant t \leqslant 2 \\ 0 & \text{otherwise} \end{cases}$$

Find the probability that, on a randomly chosen day, Lucy has to wait for less than half a minute at the traffic light. **[3]**



b

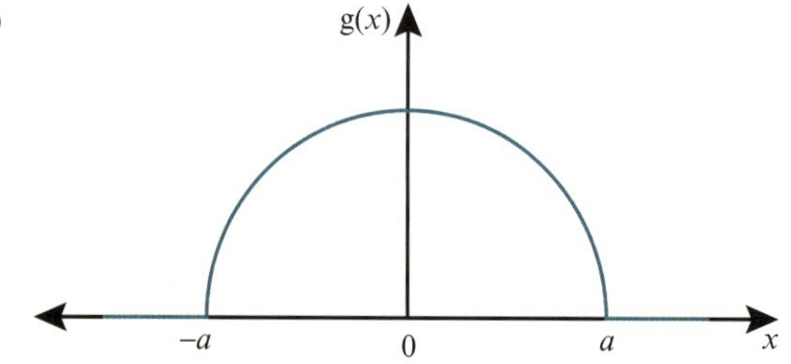


The diagram shows the graph of the probability density function, g, of a random variable, $X$. The graph of g is a semicircle with centre (0, 0) and radius $a$. Elsewhere $g(x) = 0$.

i Find the value of $a$. **[2]**

ii State the value of $E(X)$. **[1]**

iii Given that $P(X < -c) = 0.2$, find $P(X < c)$. **[2]**

*Cambridge International AS & A Level Mathematics 9709 Paper 72 Q3 November 2014*

8

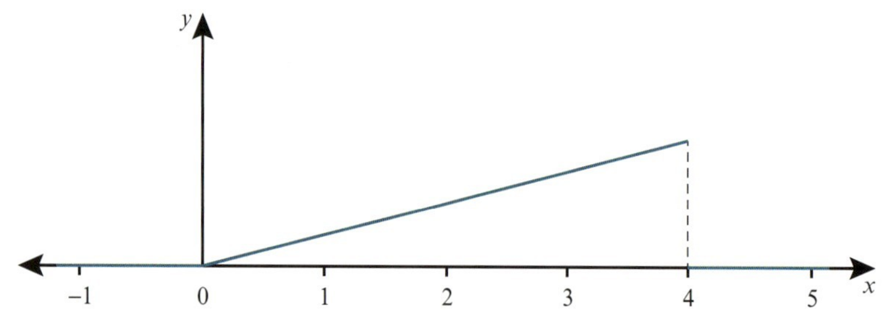


A random variable, $X$, takes values between 0 and 4 only and has probability density function as shown in the diagram. Calculate the median of $X$. **[3]**

*Cambridge International AS & A Level Mathematics 9709 Paper 73 Q2 June 2014*

9 The lifetime, $X$ years, of a certain type of battery has probability density function given by:

$$f(x) = \begin{cases} \dfrac{k}{x^2} & 1 \leqslant x \leqslant a \\ 0 & \text{otherwise} \end{cases}$$

where $k$ and $a$ are positive constants.

i State what the value of $a$ represents in this context. [1]

ii Show that $k = \dfrac{a}{a-1}$. [3]

iii Experience has shown that the longest that any battery of this type lasts is 2.5 years. Find the mean lifetime of batteries of this type. [3]

*Cambridge International AS & A Level Mathematics 9709 Paper 73 Q5 June 2014*

10

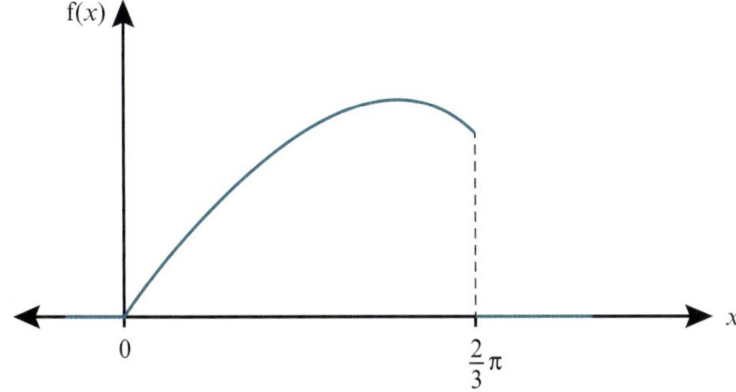


A random variable, $X$, has probability density function given by:

$$f(x) = \begin{cases} k \sin x & 0 \leqslant x \leqslant \dfrac{2}{3}\pi \\ 0 & \text{otherwise} \end{cases}$$

where $k$ is a constant, as shown in the diagram.

i Show that $k = \dfrac{2}{3}$. [2]

ii Show that the median of $X$ is 1.32, correct to 3 significant figures. [4]

iii Find $E(X)$. [4]

*Cambridge International AS & A Level Mathematics 9709 Paper 73 Q7 June 2012*

11 A random variable, $X$, has probability density function given by:

$$f(x) = \begin{cases} k(1-x) & -1 \leqslant x \leqslant 1, \\ 0 & \text{otherwise} \end{cases}$$

where $k$ is a constant.

i Show that $k = \dfrac{1}{2}$. [2]

ii Find $P\left(X > \dfrac{1}{2}\right)$. [1]

iii Find the mean of $X$. [3]

iv Find $a$ such that $P(X < a) = \dfrac{1}{4}$. [3]

*Cambridge International AS & A Level Mathematics 9709 Paper 72 Q7 June 2011*

 PS

**12** The time in minutes taken by candidates to answer a question in an examination has probability density function given by:

$$f(t) = \begin{cases} k(6t - t^2) & 3 \leqslant t \leqslant 6 \\ 0 & \text{otherwise} \end{cases}$$

where $k$ is a constant.

**i** Show that $k = \frac{1}{18}$. **[3]**

**ii** Find the mean time. **[3]**

**iii** Find the probability that a candidate, chosen at random, takes longer than 5 minutes to answer the question. **[2]**

**iv** Is the upper quartile of the times greater than 5 minutes, equal to 5 minutes or less than 5 minutes? Give a reason for your answer. **[2]**

*Cambridge International AS & A Level Mathematics 9709 Paper 71 Q5 June 2009*

# Chapter 5
# Sampling

**In this chapter you will learn how to:**

- understand the distinction between a sample and a population
- use random numbers and appreciate the necessity for choosing random samples
- explain why a sampling method may be unsatisfactory
- recognise that a sample mean can be regarded as a random variable, and use the facts that $\mathrm{E}(\bar{X}) = \mu$ and $\mathrm{Var}(\bar{X}) = \frac{\sigma^2}{n}$
- use the fact that $\bar{X}$ has a normal distribution if $X$ has a normal distribution
- use the central limit theorem where appropriate.

## PREREQUISITE KNOWLEDGE

| Where it comes from | What you should be able to do | Check your skills |
|---|---|---|
| Probability & Statistics 1, Chapter 8 | Calculate probabilities using the normal distribution. | **1** The random variable $X \sim N(22, 4^2)$. Find: **a** $P(X < 27)$ **b** $P(20 < X \leqslant 23)$ **2** The random variable $X \sim N(35, 12)$. Find: **a** $P(X > 32)$ **b** $P(X \leqslant 30)$ |

## Why do we study sampling?

In our study of statistics so far, work has focused on information collected from a **population**. Where a **sample** has been referred to, such as in hypothesis testing, this has been to find out if the information collected from the sample data fits with the population parameters.

'Population' refers to the complete collection of items of interest. It can be as diverse as a population of people or a population of fish or a population of LED lights.



A 'sample' is a part of a population. A sample can be small or large. For example, Genie has a collection of 2984 buttons. The whole population is the 2984 buttons. We may want to investigate the average size of the buttons in Genie's collection. Measuring the button size of anything fewer than all 2984 buttons, whether it is 10 buttons, 100 buttons or 1000 buttons, is to collect information from a sample of the population.

Sample data are used to test theories or to hypothesise about a population. You may think it is preferable to make decisions based on all possible data, but it is only in limited circumstances that a complete data set is sought. Data can be collected from the whole population: census data is an example. A government census collects a complete set of data from every member of the population about all aspects of life to monitor demographic and lifestyle changes. This enables the government to plan for the future needs of the population for housing, education, health, and so on. Census data are expensive and time-consuming to collect, process and analyse.

Sometimes it is not sensible to find information from every member of the population, such as in impact testing of motorcycle helmets. If all motorcycle helmets were impact tested, they would all be tested to destruction and the motorcycle helmets would no longer be usable. In these situations, a study of samples from the population is used and not the whole population. While it is true that the more information you know the more reliable your conclusions, there are reasons why we prefer to use sample data, for example:

- it is less expensive as fewer resources are needed to gather the data
- there will be fewer recording errors as there is a smaller amount of data to process
- it takes less time to collect and analyse the data
- it is more practical, as fewer items of data are needed.

Large sets of data are more reliable than small sets of data. A reasonably large sample chosen at random from a large population will generally be representative of the characteristics of the population. These two statements allow you to make valid conclusions from a sample of data. They form the basis of the **central limit theorem**.

To address the question of how reliable parameters, such as the mean of a sample, are as an estimate of the mean of the whole population, it is necessary to consider the nature of the sample; how representative it is of the whole population, the size of the sample and the variability of the population data.

**FAST FORWARD**

You will learn more about the central limit theorem later in this chapter.

**DID YOU KNOW?**

The volume of data collected by a government national population census led to increasingly sophisticated ways of collating and analysing the data. The 1880 United States of America population census used punched cards and mechanical means to collate the data. Without these mechanical means, there would not have been time enough to analyse the data collected in 1880 before the next census in 1890.

The first modern national population census in the United Kingdom was in 1841. The 1961 United Kingdom population census was the first to be analysed using a computer. Most countries collect census data every 10 years; however, in 1966 the United Kingdom government carried out its one and only mid-term 10% sample census. 'Sample census' appears to be a contradiction, since a sample is part of a population and a census is all of a population; however, the purpose of this mid-term sample census was to trial alternative methods of data collection.

**EXPLORE 5.1**

- Discuss how the UK government could choose a 10% 'sample census'.
- Which of your options do you think would give the most representative sample?
- Discuss why a sample needs to be representative if results are to be reliable.

## 5.1 Introduction to sampling

The best possible sample should have all the characteristics of the whole population. Ideally, it should have properties and characteristics identical to those of the whole population, but this is almost impossible to achieve. If one or more characteristics of a population are over- or under-represented, then you may introduce **bias**. If bias occurs, then one or more parts of the population are favoured over others, making the sample unrepresentative.

**KEY POINT 5.1**

A population is the complete set of items of interest.

A sample is part of a population.

A sample should be representative of the population.

A sampling technique that results in an unrepresentative sample is said to be biased.

**Random sampling** is the most basic of all sampling techniques. Here is a definition of a random sample:

> *A random sample of size n is a sample selected in such a way that all possible samples of size n that you could select from the population have the same chance of being chosen.*

To have the option of selecting all possible samples, every member of the population must have an equal chance of selection, and so we need a complete list of the whole population. However, this may not always be possible. For example, if the population was all the fish in the sea, or all the stars in the sky, it would be impossible to list them all. Sometimes a complete list simply does not exist; for example, of all left-handed males in the world.



If a complete list of the whole population does not exist, then some members of the population cannot be selected, and the sample selected may be unrepresentative of the whole population. This is **selection bias**, the unintentional selection of one group or outcome of a population over potential other groups or outcomes of the population.

**KEY POINT 5.2**

Random sampling means that every possible sample of a given size is equally likely to be chosen.

A sampling technique can be described as biased; the resulting sample is described as unrepresentative.

**EXPLORE 5.2**

Depending on the way data are collected, there are other ways a sample can be unrepresentative.

Discuss the following scenarios on the topic of voting habits. Give reasons to explain why the sampling technique may be biased.

1 Suppose you wish to find out views about voting in elections. Why will the sampling method be biased if you choose only people you meet on a Monday afternoon outside a school? What other ways can you think of to introduce bias in collecting the data? How could you choose a sample to try to avoid bias?

2 Suppose one of the questions in your survey is: 'Do you agree that everyone who can vote should vote?'. Why does beginning a question like this invite you to agree? Can you rewrite the question so that it does not influence your answer?

3 Suppose you ask the question, 'Have you ever donated money to a political party?'. Is this actually useful information with respect to voting habits? How do you ensure the data collected are relevant?

4 At an election rally, everyone is given a glow stick. The company making the glow sticks advertises that the lights last for an average of 3 hours. Do you think the company tests every glow stick? Why not? If the manufacturer made 5000 glow sticks on one day, why is there bias in testing just the first 50 the machine made that day? How would you choose 50 glow sticks to test from a batch of 5000?

5 You find out opinions from 20 people. Is this enough information? How do you know when you have collected enough data?

## Use of random numbers

A random sample gives every member of the population the same chance of being selected. The complete list of all items or people in the population is the **sampling frame**. For a small population, the method of generating a random sample could be as basic as giving every member of the population a number, placing numbered discs in a box and choosing discs at random.

You can always generate a random sample, regardless of population size, using random numbers.



**DID YOU KNOW?**

Early methods of generating values entirely by chance, such as rolling a die or flipping a coin, are still used today when playing games, but these methods produce random numbers too slowly for statisticians.

Early random number tables generated values in different ways, such as by choosing numbers 'at random' from census data or from logarithmic tables. One of the earliest examples of tables of random numbers was published in 1927 by Cambridge University Press; the table contained 41 600 digits.

Today, with the advent of information technology, organisations collecting data often prefer to use computational random number generators rather than tables of random numbers. An early example of this occurred in the 1940s when the RAND Corporation generated random numbers using a simulation of a roulette wheel attached to a computer. Calculators use RAND or RND or RAN# to generate random numbers.

Random number generators have many applications; for example, statistical sampling, computer simulation and cryptography. In the 1950s, a hardware random number generator named ERNIE was first used to draw British premium bond numbers.

**EXPLORE 5.3**

Microsoft Excel has a function to produce random numbers. The function is simply:

**=RAND()**

In Excel, type =**RAND()** into a cell, then copy the formula down the column and across the row to produce your own table of random numbers between 0 and 1.

To generate a different range of random numbers, such as between **1** and **250**, enter the following formula:

**=INT(250*RAND())+1**

The **INT** eliminates the digits after the decimal point, the **250*** creates the range to be covered, and the +**1** sets the lowest number in the range.

Adapt the formula to generate your own page of random numbers for a different range of values.

The following process outlines one way to use a table of random numbers to find a random sample.

Suppose you want to choose a sample of 20 people from a population of 500. First, number each person from 000 to 499; next, decide on a starting point in a table of random numbers such as the one following.



| | | | | |
|---|---|---|---|---|
| 68236 | 35335 | 71329 | 96803 | 24413 |
| 62385 | 36545 | 59305 | 59948 | 17232 |
| 64058 | 80195 | 30914 | 16664 | 50818 |
| 64822 | 28554 | 90952 | 64984 | 92295 |
| 17716 | 22164 | 05161 | 04412 | 59002 |
| 03928 | 22379 | 92325 | 79920 | 99070 |
| 11021 | 08533 | 83855 | 37723 | 77339 |
| 01830 | 28554 | 86787 | 90447 | 54796 |
| 36782 | 73208 | 93548 | 77405 | 58355 |
| 58158 | 45059 | 83980 | 40176 | 40737 |
| 91239 | 10532 | 27993 | 11516 | 61327 |
| 27073 | 98804 | 60544 | 12133 | 01422 |
| 81501 | 00633 | 62681 | 84319 | 03374 |
| 64374 | 26598 | 54466 | 94768 | 19144 |
| 29896 | 26739 | 30871 | 29795 | 13472 |
| 38996 | 72151 | 65746 | 16513 | 62796 |
| 73936 | 81751 | 00149 | 99126 | 23117 |
| 18795 | 93118 | 84105 | 18307 | 49807 |
| 76816 | 99822 | 92314 | 45035 | 43490 |
| 12091 | 60413 | 90467 | 42457 | 50490 |

Starting at the top of the second column and reading the first three digits of each set of five digits, going down the column produces the following list:

353 365 801 285 221 223 085 285 732 ....

The number 285 appears more than once, so it is ignored the second time and any subsequent times it appears.

One way to deal with numbers greater than 499 is to ignore them: we can move on to the next random number until 20 have been selected. Alternatively, in this example of a population of 500, 500 could be subtracted from each number over 499 and the new value used as the random number; for example, 801 becomes 301, 732 becomes 232 and so on.

We could start at any place in the table and read across rows or down columns. Although the random numbers in this table are in blocks of five digits, with five blocks in each cell of the table, this is simply to make it easier for you to keep track when using them. The digits can be used as if they had been written in one long, continuous line. When you reach the end of the row or column and need more values, you simply move to the next row or column, respectively.

**KEY POINT 5.3**

Random numbers can be used to generate a sample in which you have no control over the selection. However, bear in mind that using a random sampling method does not guarantee that the resulting sample will be representative of the whole population.

**EXPLORE 5.4**

Use Microsoft Excel, or the random number function on your calculator, to produce five random single-digit numbers. Work out the mean of the five random numbers. Repeat this process eight times.

Plot the mean values obtained as a frequency graph, together with the results from each member of your teaching group. What do you notice about the shape of your graph?

What do you think would happen to the shape of the graph if you repeated this process many more times?

What do you think may happen if you plot the means of 20 single-digit numbers chosen at random instead of five random numbers? Try it and see.

We will further explore these results later in this chapter.



## EXERCISE 5A

1 Use the random number table from earlier in this chapter to identify the first six random numbers, using the following criteria:

 a three-digit numbers starting at the top of the third column and reading across

 b three-digit numbers starting at line six of the first column and reading across.

2 Forid uses the random number table from earlier in this chapter to identify the first four two-digit random numbers, starting at the top of the second column and using only that column. Explain how Forid obtains each of the following sets of random numbers.

 a 35 33 53 65 b 35 36 80 28 c 33 82 22 02

3 Niheda wishes to choose a representative sample of six employees from the 78 employees at her place of work.

a Niheda considers taking as her sample the first six people arriving at work one morning. Give two reasons why this method is unsatisfactory.

b Niheda decides to use the following method to choose her sample. She numbers each employee at her place of work and generates the following random numbers on her calculator:

642 784 034 796 313 215 950 850 565 013 311 170 929

From these random numbers, she chooses employees 40 47 63 32 59 and 8. Explain how she chose these employees.

4 The manufacturer of a new chocolate bar wishes to find out what people think of it. The manufacturer decides to interview a sample of people. Describe the bias in the method used to select each of the following samples.

a A sample of people who have just bought the chocolate bar.

b A sample of people aged between 25 and 29.

c The first 20 males shopping at a store where the chocolate bar is sold.

5 Milek wishes to choose a sample of four students from a class of 16 students. The students are numbered from 3 to 18, inclusive. Milek throws three fair dice and adds the scores. Explain why this method of choosing the sample is biased.

6 Describe briefly how to use random numbers to choose a sample of 50 employees from a company with 712 employees.



## 5.2 The distribution of sample means

Different samples of data chosen from the same population will most likely have different, but not necessarily dissimilar, means.

The **sample mean** is the mean of all the items in your chosen sample.

The sample size is the number of items you choose to be in your sample.

You can explore the distribution of sample means using any distribution, discrete or continuous, normal or otherwise, so long as it has a defined mean value.

For example, suppose you spin a fair four-sided spinner, numbered 1, 1, 2 and 4, a number of times. With a sample size of five, you could get the following outcomes:

1 1 4 2 4 or 1 1 1 4 1 or 4 1 2 2 1 or…

There are too many possibilities to list.

The sample means for each of the samples presented are $\frac{12}{5} = 2.4$, $\frac{8}{5} = 1.6$ and $\frac{10}{5} = 2$, respectively. It would take a very long time to list all possible samples and work out each sample mean. If we did, we could create a table showing the probability distribution of the sample mean and create a graph to show the distribution of the sample means.

To show how this works we can begin with the simplest sample size, 1, and explore the probability distribution of the means of increasing sample sizes using this fair four-sided spinner.

Let the random variable $X$ be 'the score on the spinner when it is spun'.

When we spin the spinner, it is equally likely to land on each side. With a sample size of 1 the mean of the sample is the same as the score.

To distinguish between the probability distribution of scores and the probability distribution of sample means we use $\bar{X}$ to represent the random variable of sample means. To follow the explanation it is easier to refer to the distribution of sample size 1 as $\bar{X}(1)$.

The probability distribution of $\bar{X}(1)$ is:

| **Sample mean, $x$** | 1 | 2 | 4 |
|---|---|---|---|
| **$P(\bar{X}(1)) = x$** | $\frac{1}{2}$ | $\frac{1}{4}$ | $\frac{1}{4}$ |

The following figure shows the graph of the probability distribution of the sample means of size 1.

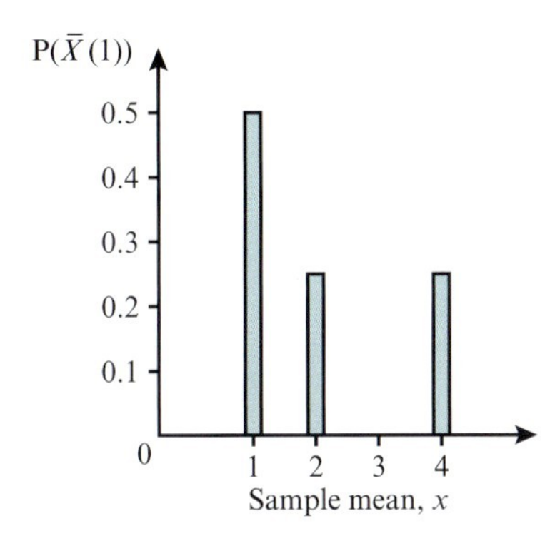


$$E(\bar{X}(1)) = \left(1 \times \frac{1}{2}\right) + \left(2 \times \frac{1}{4}\right) + \left(4 \times \frac{1}{4}\right)$$
$$= 2$$
$$\text{Var}(\bar{X}(1)) = \left(1^2 \times \frac{1}{2}\right) + \left(2^2 \times \frac{1}{4}\right) + \left(4^2 \times \frac{1}{4}\right) - (E(X))^2$$
$$= 5.5 - 2^2$$
$$= 1.5$$

Suppose we now choose random samples of size 2. If $X_1$ is the score from the first spin and $X_2$ the score from the second spin, then we can draw a table to show all possible sample means.

|  | $X_1$ |  |  |  |  |
|---|---|---|---|---|---|
|  |  | 1 | 1 | 2 | 4 |
| $X_2$ | 1 | 1 | 1 | $1\frac{1}{2}$ | $2\frac{1}{2}$ |
|  | 1 | 1 | 1 | $1\frac{1}{2}$ | $2\frac{1}{2}$ |
|  | 2 | $1\frac{1}{2}$ | $1\frac{1}{2}$ | 2 | 3 |
|  | 4 | $2\frac{1}{2}$ | $2\frac{1}{2}$ | 3 | 4 |

From this we can find the probability distribution of the sample means of size 2, $\bar{X}(2)$. For example, there are 16 possible sample means and the sample mean $1\frac{1}{2}$ appears four times in the table; hence, $P(\bar{X}(2) = 1\frac{1}{2}) = \frac{4}{16} = \frac{1}{4}$.

| **Sample mean, $x$** | 1 | $1\frac{1}{2}$ | 2 | $2\frac{1}{2}$ | 3 | 4 |
|---|---|---|---|---|---|---|
| **$P(\bar{X}(2)) = x$** | $\frac{1}{4}$ | $\frac{1}{4}$ | $\frac{1}{16}$ | $\frac{1}{4}$ | $\frac{1}{8}$ | $\frac{1}{16}$ |

This probability distribution is the distribution of the sample means of size 2. The following diagram shows the graph of the probability distribution of the sample means of size 2.

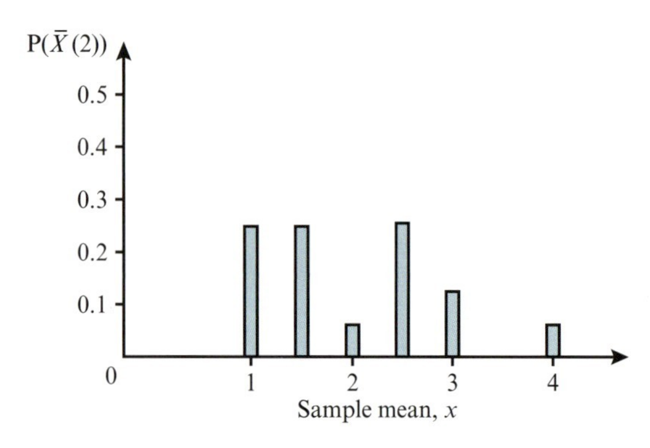


$$\mathrm{E}(\bar{X}(2)) = \left(1 \times \frac{1}{4}\right) + \left(1\frac{1}{2} \times \frac{1}{4}\right) + \left(2 \times \frac{1}{16}\right) + \left(2\frac{1}{2} \times \frac{1}{4}\right) + \left(3 \times \frac{1}{8}\right) + \left(4 \times \frac{1}{16}\right)$$
$$= 2$$

$$\mathrm{Var}(\bar{X}(2)) = \left(1^2 \times \frac{1}{4}\right) + \left(\left(1\frac{1}{2}\right)^2 \times \frac{1}{4}\right) + \left(2^2 \times \frac{1}{16}\right) + \left(\left(2\frac{1}{2}\right)^2 \times \frac{1}{4}\right)$$
$$+ \left(3^2 \times \frac{1}{8}\right) + \left(4^2 \times \frac{1}{16}\right) - 2^2$$
$$= 4.75 - 4 = 0.75$$

Note that $\mathrm{E}(\bar{X}(1)) = \mathrm{E}(\bar{X}(2))$, whereas $\mathrm{Var}(\bar{X}(1)) \neq \mathrm{Var}(\bar{X}(2))$.

In fact, $\mathrm{Var}(\bar{X}(2)) = \frac{1}{2}\mathrm{Var}(\bar{X}(1))$.

To confirm these results will always work, we can explore what happens when we take a sample size of 3.



The probability distribution of the sample mean scores for $\bar{X}(3)$ is shown in the following table.

| **Sample mean, $x$** | 1 | $\frac{4}{3}$ | $\frac{5}{3}$ | 2 | $\frac{7}{3}$ | $\frac{8}{3}$ | 3 | $\frac{10}{3}$ | 4 |
|---|---|---|---|---|---|---|---|---|---|
| **$\mathrm{P}(\bar{X}(3)) = x$** | $\frac{1}{8}$ | $\frac{3}{16}$ | $\frac{3}{32}$ | $\frac{13}{64}$ | $\frac{3}{16}$ | $\frac{3}{64}$ | $\frac{3}{32}$ | $\frac{3}{64}$ | $\frac{1}{64}$ |

**EXPLORE 5.5**

See if you can verify the probabilities of all other scores in the distribution table for samples of size 2.

The following diagram shows the graph of the probability distribution of the sample means of size 3.

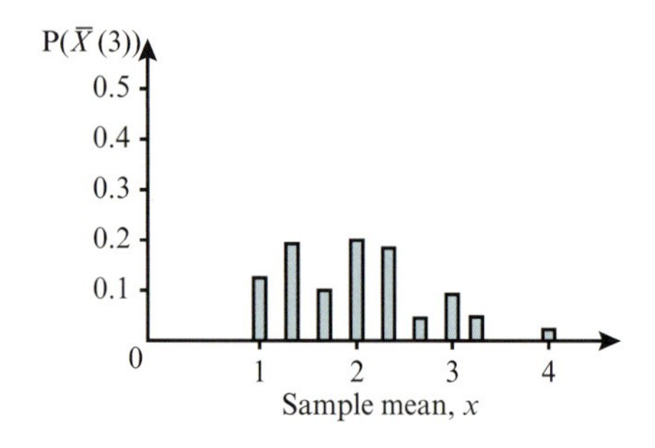


**REWIND**

We can explore these results using our knowledge of linear combinations of random variables from Chapter 3. From Chapter 3 we know that:

$\mathrm{E}(X_1 + X_2) = \mathrm{E}(X_1) + \mathrm{E}(X_2)$ and

$\mathrm{Var}(X_1 + X_2) = \mathrm{Var}(X_1) + \mathrm{Var}(X_2)$.

Using these results to check our findings from the table, we find:

$$\mathrm{E}(\bar{X}(2)) = \mathrm{E}\left(\frac{1}{2}(X_1 + X_2)\right)$$
$$= \mathrm{E}\left(\frac{1}{2}X_1 + \frac{1}{2}X_2\right)$$
$$= \frac{1}{2}\mathrm{E}(X) + \frac{1}{2}\mathrm{E}(X)$$
$$= \mathrm{E}(X) = 2 \text{ (as before)}$$
$$\mathrm{Var}(\bar{X}(2)) = \mathrm{Var}\left(\frac{1}{2}(X_1 + X_2)\right)$$
$$= \mathrm{Var}\left(\frac{1}{2}X_1 + \frac{1}{2}X_2\right)$$
$$= \frac{1}{2^2}\mathrm{Var}(X) + \frac{1}{2^2}\mathrm{Var}(X)$$
$$= \frac{1}{2}\mathrm{Var}(X) = \frac{1}{2} \times 1.5$$
$$= 0.75 \text{ (as before)}$$

**REWIND**

We can use our knowledge of permutations and combinations, from Probability & Statistics 1 Coursebook, Chapter 5, to list all possible outcomes for a sample size of 3.

For example, for a sample of size 3, a mean score $\frac{7}{3}$ can happen in the following six ways:

1 2 4  1 4 2  2 1 4  2 4 1  4 1 2  4 2 1

Each arrangement has a probability of $\frac{1}{2} \times \frac{1}{4} \times \frac{1}{4} = \frac{1}{32}$. Hence, the probability of obtaining a mean score $\frac{7}{3}$ is $\frac{6}{32} = \frac{3}{16}$.

For a sample of size 3, a mean score 2 can happen in the following four ways:

1 1 4  1 4 1  4 1 1  2 2 2

And the probability of a mean score of 2 is:

$$3\left(\frac{1}{2} \times \frac{1}{2} \times \frac{1}{4}\right) + \frac{1}{4} \times \frac{1}{4} \times \frac{1}{4} = \frac{3}{16} + \frac{1}{64}$$
$$= \frac{13}{64}$$

$$E(\bar{X}(3)) = \left(1\times\frac{1}{8}\right) + \left(\frac{4}{3}\times\frac{3}{16}\right) + \left(\frac{5}{3}\times\frac{3}{32}\right) + \left(2\times\frac{13}{64}\right) + \left(\frac{7}{3}\times\frac{3}{16}\right)$$
$$+ \left(\frac{8}{3}\times\frac{3}{64}\right) + \left(3\times\frac{3}{32}\right) + \left(\frac{10}{3}\times\frac{3}{64}\right) + \left(4\times\frac{1}{64}\right)$$
$$= \frac{1}{8}+\frac{1}{4}+\frac{5}{32}+\frac{13}{32}+\frac{7}{16}+\frac{1}{8}+\frac{9}{32}+\frac{5}{32}+\frac{1}{16}$$
$$= \frac{64}{32} = 2$$

$$\text{Var}(\bar{X}(3)) = \left(1^2\times\frac{1}{8}\right) + \left(\left(\frac{4}{3}\right)^2\times\frac{3}{16}\right) + \left(\left(\frac{5}{3}\right)^2\times\frac{3}{32}\right) + \left(2^2\times\frac{13}{64}\right) + \left(\left(\frac{7}{3}\right)^2\times\frac{3}{16}\right)$$
$$+ \left(\left(\frac{8}{3}\right)^2\times\frac{3}{64}\right) + \left(3^2\times\frac{3}{32}\right) + \left(\left(\frac{10}{3}\right)^2\times\frac{3}{64}\right) + \left(4^2\times\frac{1}{64}\right) - 2^2$$
$$= \frac{1}{8}+\frac{1}{3}+\frac{25}{96}+\frac{13}{16}+\frac{49}{48}+\frac{1}{3}+\frac{27}{32}+\frac{25}{48}+\frac{1}{4} - 2^2$$
$$= 4\tfrac{1}{2} - 4 = \frac{1}{2}$$

Alternatively, using linear combinations of random variables:

If $X_1$ is the score from the first spin, $X_2$ the score from the second spin, and $X_3$ the score from the third spin, then:

$$E(\bar{X}(3)) = E\left(\frac{1}{3}X_1 + \frac{1}{3}X_2 + \frac{1}{3}X_3\right) = E(X) = 2 \text{ (as before)}$$

$$\text{Var}(\bar{X}(3)) = \text{Var}\left(\frac{1}{3}X_1 + \frac{1}{3}X_2 + \frac{1}{3}X_3\right)$$
$$= \frac{1}{3^2}\text{Var}(X_1) + \frac{1}{3^2}\text{Var}(X_2) + \frac{1}{3^2}\text{Var}(X_3)$$
$$= \frac{1}{3}\text{Var}(X) = \frac{1}{3}\times 1.5 = \frac{1}{2} \text{ (as before)}$$

**EXPLORE 5.6**

Are you able to use your knowledge of linear combinations of random variables to work out the results $E(\bar{X}(4))$ and $\text{Var}(\bar{X}(4))$ for a sample of size 4?

What about our original sample size of 5? Can you find $E(\bar{X}(5))$ and $\text{Var}(\bar{X}(5))$ without having to list all possible samples?

Can you extend your results to sample size $n$?

What do you think the graph of the sample means will look like as the sample size $n$ increases?

**KEY POINT 5.4**

If you take many samples and calculate the mean of each sample, these means have a distribution called the distribution of the sample mean. A sample mean can be regarded as a random variable.

If a random sample consists of $n$ observations of a random variable $X$ and the mean $\bar{X}$ is found, then:

$$E(\bar{X}(n)) = \mu \text{ where } \mu = E(X) \text{ and } \text{Var}(\bar{X}(n)) = \frac{\sigma^2}{n} \text{ where } \sigma^2 = \text{Var}(X).$$

**WORKED EXAMPLE 5.1**

**a** Show that for samples of size 1 drawn from a fair six-sided die numbered 1, 2, 3, 4, 5 and 6, $\mathrm{E}(\bar{X}(1)) = 3\frac{1}{2}$ and $\mathrm{Var}(\bar{X}(1)) = \frac{35}{12}$.

**b** Work out $\mathrm{E}(\bar{X}(2))$ and $\mathrm{Var}(\bar{X}(2))$.

**Answer**

**a** $\mathrm{E}(\bar{X}(1)) = \left(1\times\frac{1}{6}\right)+\left(2\times\frac{1}{6}\right)+\left(3\times\frac{1}{6}\right)+\left(4\times\frac{1}{6}\right)+\left(5\times\frac{1}{6}\right)+\left(6\times\frac{1}{6}\right)$

You may choose to draw a probability distribution table.

$= \frac{21}{6} = 3\frac{1}{2}$

$\mathrm{Var}(\bar{X}(1)) = \left(1^2\times\frac{1}{6}\right)+\left(2^2\times\frac{1}{6}\right)+\left(3^2\times\frac{1}{6}\right)+\left(4^2\times\frac{1}{6}\right)$

$+\left(5^2\times\frac{1}{6}\right)+\left(6^2\times\frac{1}{6}\right)-\left(3\frac{1}{2}\right)^2$

$= \frac{91}{6} - \frac{49}{4} = \frac{35}{12}$

**b** $\mathrm{E}(\bar{X}(2)) = \frac{1}{2}\mathrm{E}(X) + \frac{1}{2}\mathrm{E}(X) = \mathrm{E}(X) = 3\frac{1}{2}$

You can use expectation algebra, as you have found $\mathrm{E}(X)$ and $\mathrm{Var}(X)$.

$\mathrm{Var}(\bar{X}(2)) = \frac{1}{2^2}\mathrm{Var}(X) + \frac{1}{2^2}\mathrm{Var}(X) = \frac{1}{2}\mathrm{Var}(X) = \frac{1}{2}\times\frac{35}{12} = \frac{35}{24}$



## The central limit theorem

We have now found that the means of random samples of size $n$ from a population with mean $\mu$ and variance $\sigma^2$ will have a distribution with mean $\mu$ and variance $\frac{\sigma^2}{n}$, but what sort of distribution will it be?

Below are the graphs of the probability distributions for sample means of size 1, 2 and 3 for the spinner numbered 1, 1, 2 and 4.

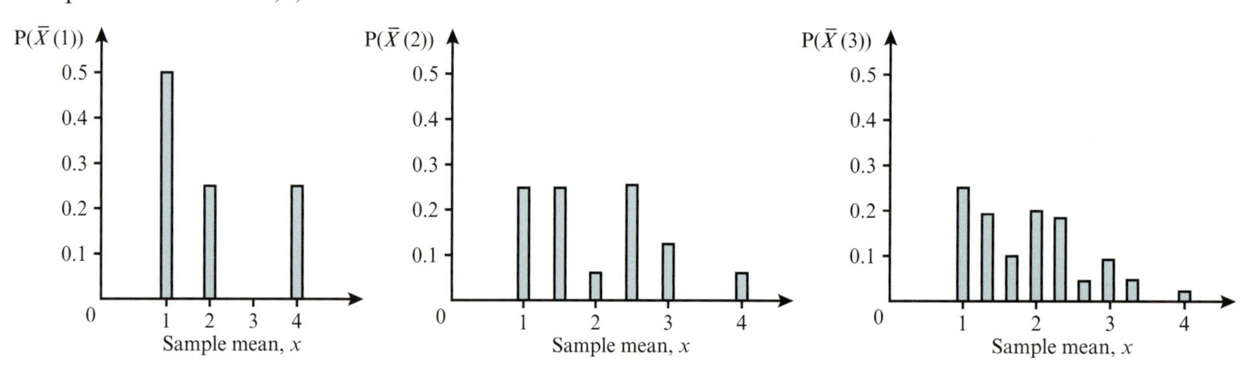


We can see the shape of the distribution changes as $n$ increases; this tells us the distribution of sample means does not depend on the shape of the original distribution.

To examine the shape of the probability distribution of sample means as $n$ increases, let us explore an example using a more familiar object, the sample mean of scores on an ordinary fair die, numbered 1, 2, 3, 4, 5 and 6. The following graph shows the probability distribution of the sample means of size 1.

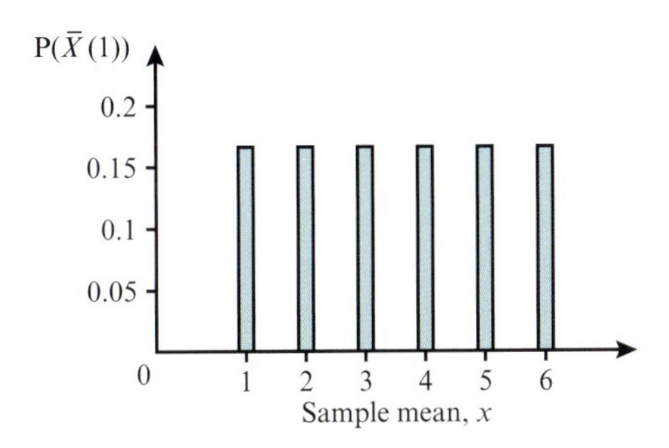


For sample size of 1, the probability distribution graph is uniform; each score has probability $\frac{1}{6}$.

The following graph shows the probability distribution of the sample means of size 2.

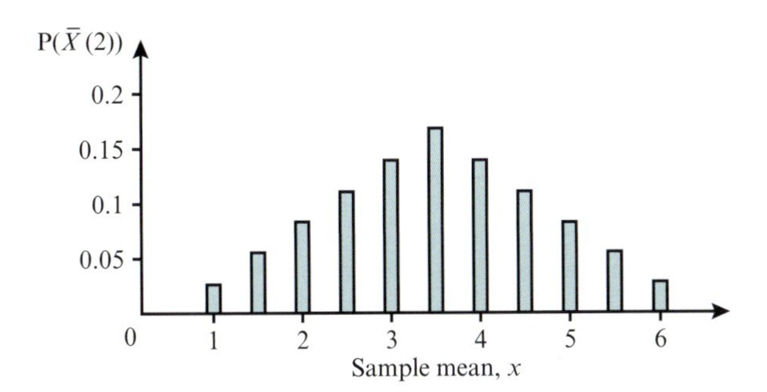


For sample of size 2, the probability distribution graph is symmetrical about the mean value.

If we draw probability distribution graphs for larger sample sizes; for example, samples of size 6 and 10, shown on the following graphs, we can see the distribution of sample means increasingly begins to resemble a normal distribution.

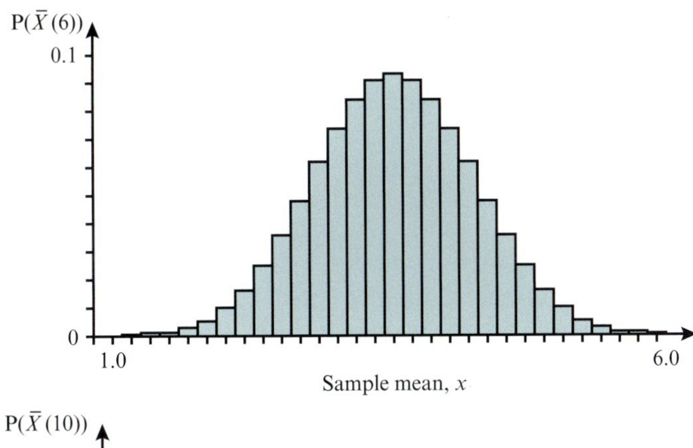


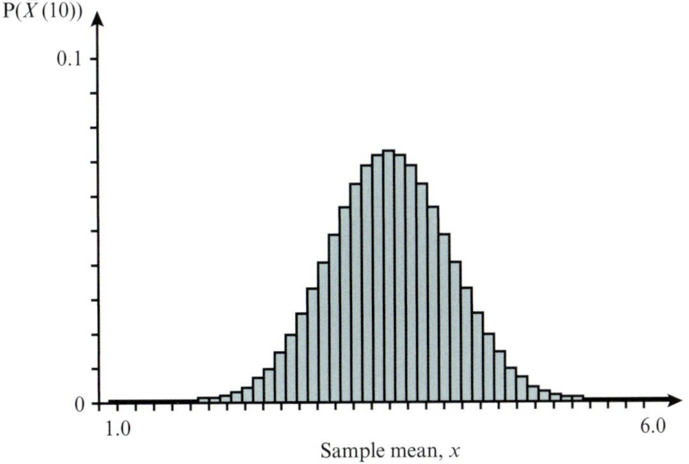

For a fair ordinary die, and where the sample size is 1, the original distribution is uniform (rectangular). From sample of size 2 onwards, the graphs of the probability distributions of sample means show the peak of the graph at the mean of the original distribution.

As the sample size increases, the probability of getting a sample mean further away from the actual mean of the distribution, such as the mean of samples of size 1, becomes smaller and smaller. Hence, the variance of the distribution of the sample mean becomes smaller as $n$ becomes larger.

**EXPLORE 5.7**

You can create probability distributions graphs for different sample sizes. Search 'dice experiment' for a suitable program to use.

**EXPLORE 5.8**

Look back at your graphs generated from means of samples of single-digit random numbers from Explore 5.4. What conclusions can you now draw from these graphs?

**KEY POINT 5.5**



For large sample sizes, the distribution of a sample mean is approximately normal. This normal distribution will have mean $\mu$ and variance $\frac{\sigma^2}{n}$.

This result is true for all distributions of sample means, regardless of whether the underlying population is normal. This is the fundamental property of the central limit theorem.

The central limit theorem (CLT) states that, provided $n$ is large, the distribution of sample means of size $n$ is:

$$\bar{X}(n) \sim \mathrm{N}\left(\mu, \frac{\sigma^2}{n}\right),$$ where the original population has mean $\mu$ and variance $\sigma^2$.

The value of sample size $n$ required for the central limit theorem to be a good approximation depends on the original population distribution. We need to decide how large is sufficiently large a value of $n$ to use the central limit theorem as a good approximation. This depends on the distribution of the original population. If the original population is approximately normal, then the distribution of sample means for a low value of $n$ is sufficient. However, if the original population does not display any features of a normal distribution, then the value of $n$ will need to be large. For any population, the central limit theorem can be used for sample size $n > 50$.

It follows that if the original distribution is normal, $X \sim \mathrm{N}(\mu, \sigma^2)$, then the distribution of sample means from a normal distribution must also be a normal distribution since $\mathrm{E}(\bar{X}(n)) = \mu$ and $\mathrm{Var}(\bar{X}(n)) = \frac{\sigma^2}{n}$. As $n$ increases, the shape of the distribution of sample means becomes more peaked and centred around $\mu$, as shown in the following diagram.

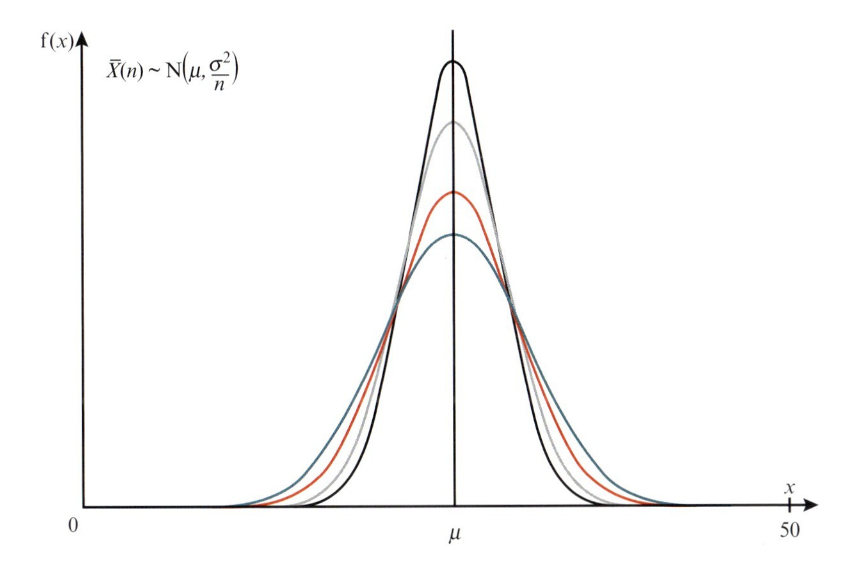


**KEY POINT 5.6**

The central limit theorem is important when samples of data are being explored, because the distribution of means of samples is approximately normal even when the parent population is not normal. The central limit theorem allows the use of the normal distribution to make statistical judgements from sample data from any distribution.



**WORKED EXAMPLE 5.2**

The masses of a variety of pears are normally distributed with mean 45 g and variance 52 g$^2$. The pears are packed in bags of six. Find the percentage of bags of pears with a total mass of more than 300 g.

**Answer 1**

$$1 - \Phi\left(\frac{\frac{300}{6} - 45}{\sqrt{\frac{52}{6}}}\right) = 1 - \Phi(1.698)$$

$$= 1 - 0.9553 = 0.0447 = 4.47\%$$

Sample mean $\bar{X} \sim N\left(45, \frac{52}{6}\right)$. In a bag with total mass 300 g, each will have an average mass of $\frac{300}{6}$. Use normal tables to calculate the probability and, hence, the percentage.

**Answer 2**

$$1 - \Phi\left(\frac{300 - 270}{\sqrt{312}}\right) = 1 - \Phi(1.698)$$

$$= 1 - 0.9553 = 0.0447 = 4.47\%$$

An alternative is to find the distribution for the bag of pears, and multiply the mean and variance by the number of pears in the bag. Then:

$X \sim N(270, 312)$

Use the normal tables, as before.

### WORKED EXAMPLE 5.3

During an exercise session, women will drink, on average, 500 ml of water with a standard deviation of 50 ml. 25 women are taking part in the exercise session. You have available 13 litres of water. What is the probability you will have sufficient water?

**Answer**

The situation described has $\mu = 500$ and $\sigma = 50$.

You do not know if the situation follows a normal distribution.

However, the distribution of sample means is normal, and its mean is the same as the population.

Use normal tables to calculate the probability.

The probability of sufficient water implies less than 13000 ml will be needed by the group of women.

There will be sufficient water if each woman drinks, on average, less than $\frac{13000}{25} = 520\text{ml}$.

$$\bar{X} \sim \mathrm{N}\left(500, \frac{50^2}{25}\right)$$

$$\mathrm{P}(\bar{X} < 520) = \mathrm{P}\left(Z < \frac{520 - 500}{\frac{50}{\sqrt{25}}}\right)$$

$$= \mathrm{P}(Z < 2) = 0.977$$

### WORKED EXAMPLE 5.4

A continuous random variable, $X$, has probability density given by:

$$\mathrm{f}(x) = \begin{cases} \frac{x}{2} & 0 \leqslant x \leqslant 2 \\ 0 & \text{otherwise} \end{cases}$$

Calculate the probability that the mean, $\bar{X}$, of a random sample of 50 observations of $X$ is less than $\frac{3}{2}$.

**Answer**

$$\text{Mean} = \int_0^2 \frac{x^2}{2}\,\mathrm{d}x = \left[\frac{x^3}{6}\right]_0^2 = \frac{4}{3}$$

First find the mean and variance of $X$.

Use the CLT to define the distribution of $\bar{X}$.

Use normal tables to calculate the probability.

$$\text{Variance} = \int_0^2 \frac{x^3}{2}\,\mathrm{d}x - \left(\frac{4}{3}\right)^2$$

$$= \left[\frac{x^4}{8}\right]_0^2 - \frac{16}{9} = \frac{2}{9}$$

$$\bar{X} \sim \mathrm{N}\left(\frac{4}{3}, \frac{2}{450}\right)$$

$\mathrm{Var}(X \text{ bar}) = \mathrm{Var}(X) \div n$

$$= \frac{2}{9} \div 50$$

$$= \frac{2}{450}$$

$$\mathrm{P}\left(\bar{X} < \frac{3}{2}\right) \approx \Phi\left(\frac{\frac{3}{2} - \frac{4}{3}}{\sqrt{\frac{2}{450}}}\right) = \Phi(2.5) = 0.9938$$

**REWIND**

We learnt how to find mean and variance in Chapter 4.

## WORKED EXAMPLE 5.5

An IT security firm detects threats to steal online data at the rate of 12.2 per day. The threats occur singly and at random. A random sample of 100 weeks is chosen. Find the probability that the average weekly number of threats detected is less than 86.

**Answer 1**

Let the random variable $T$ be the total number of threats detected in one week. Then $T \sim \text{Po}(85.4)$.

First define the distribution.

$\bar{T} \sim \text{N}\left(85.4, \frac{85.4}{100}\right)$

Use the CLT to define the distribution of $\bar{T}$.

$$\text{P}(\bar{T} < 86) = \Phi\left(\frac{\left(86 - \frac{1}{200}\right) - 85.4}{\sqrt{\frac{85.4}{100}}}\right)$$

$$= \Phi(0.6439) = 0.74$$

The continuity correction is $\frac{1}{2n} = \frac{1}{2 \times 100}$.

When working with discrete random variables, the continuity correction is $\pm\frac{1}{2n}$ not $\pm\frac{1}{2}$.

To explain why, for this example we can find the required probability using an alternative method.

**Answer 2**

Let the random variable $X$ be the number of threats over 100 weeks. Then:

$X \sim \text{Po}(8540) \approx \text{N}(8540, 8540)$

This time, we are using mean over the whole interval of 100 weeks.

$$P(X < 8600) \approx \Phi\left(\frac{8599.5 - 8540}{\sqrt{8540}}\right)$$

$$= \Phi(0.644) \approx 0.74$$

To calculate the probability using this method, we use the usual continuity correction $\pm\frac{1}{2}$, depending on the situation.

And we get the same answer as before.

In the first method, we found $P(\bar{T} < 86)$, whereas in the second method we found $\text{P}(X < 8600)$; 86 is 100 times smaller than 8600, and the continuity correction $\frac{1}{200}$ is 100 times smaller than $\frac{1}{2}$.

We learnt about the Poisson distribution in Chapter 2.



**TIP**

When using the central limit theorem for sample means size $n$ taken from a discrete distribution, such as the binomial or Poisson distributions, the continuity correction is $\pm\frac{1}{2n}$.

## WORKED EXAMPLE 5.6

The random variable $X \sim \text{B}(60, 0.25)$. The random variable $\bar{X}$ is the mean of a random sample of 50 observations of $X$. Find $\text{P}(\bar{X} \leqslant 16)$.

**Answer**

$\bar{X} \sim \text{N}\left(15, \frac{11.25}{50}\right)$

Mean $np = 60 \times 0.25 = 15$.

Variance $np(1 - p) = 60 \times 0.25 \times 0.75 = 11.25$.

$$\text{P}(\bar{X} \leqslant 16) = \Phi\left(\frac{16 + \frac{1}{100} - 15}{\sqrt{\frac{11.25}{50}}}\right) = \Phi(2.129) = 0.983$$

Use continuity correction $\frac{1}{2 \times 50}$.

## EXERCISE 5B

1 The random variable $X$ has mean 6 and variance 8. The random variable $\bar{X}$ is the mean of a random sample of 80 observations of $X$. State the approximate distribution of $\bar{X}$, giving its parameters, and find the probability that the sample mean is less than 6.4.

2 The random variable $X$ has mean 30 and variance 36. The random variable $\bar{X}$ is the mean of a random sample of 100 observations of $X$. State the approximate distribution of $\bar{X}$, giving its parameters, and find the probability that the sample mean is greater than 31.

3 The random variable $Y$ has mean 21 and standard deviation 4.2. The random variable $\bar{Y}$ is the mean of a random sample of 50 observations of $Y$. State the approximate distribution of $\bar{Y}$, giving its parameters, and work out $P(\bar{Y} < 22)$.

4 The time taken for telephone calls to a call centre to be answered is normally distributed with mean 20 seconds and standard deviation 5 seconds. Find the probability that for 16 randomly selected calls made to the centre, the mean time taken to answer the calls is less than 18 seconds.

PS 5 Ciara needs 5 kg of flour, so she buys 10 bags, each labelled as containing 500 g. Unknown to her, the bags contain, on average, 510 g with variance 120 $g^2$. What is the probability that Ciara actually buys less flour than she needs?

PS 6 The length, in cm, of an electrical component produced by a company may be considered to be a continuous random variable $X$, having probability density function as follows:

$$f(x) = \begin{cases} \frac{5}{2} & 1.8 \leqslant x \leqslant 2.2 \\ 0 & \text{otherwise} \end{cases}$$

a Calculate the probability that the mean, $\bar{X}$, of a random sample of 40 of these components is greater than 2.05 cm.

b Calculate the probability that the mean, $\bar{X}$, of a random sample of 20 of these components is less than 2.05 cm.

7 A random sample of size 60 is taken from the random variable $X$, where $X \sim B(45, 0.4)$. Given that $\bar{X}$ is the sample mean, find:

a $P(\bar{X} < 19)$

b $P(\bar{X} \leqslant 18)$

8 A random sample of size 50 is taken from random variable $X$, where $X \sim Po(2)$. Find $P(1.5 < \bar{X} \leqslant 2.2)$, where $\bar{X}$ is the sample mean.

## Checklist of learning and understanding

- 'Population' means all the items of interest within a study.
- 'Sample' describes part of a population.
- Biased sampling occurs when the sample is unrepresentative of the population.
- Random numbers can be used to generate a sample in which you have no control over the selection.
- Random sampling is a process whereby each member of the population has an equal chance of selection.
- Random sampling does not guarantee that the resulting sample will be representative of the population.
- The central limit theorem allows the use of the normal distribution to make statistical judgements from sample data from any distribution.
- For samples of size $n$ drawn from a population with mean $\mu$ and variance $\sigma^2$, the distribution of sample means $\overline{X}$ is normal and $\overline{X} \sim \mathrm{N}\left(\mu, \frac{\sigma^2}{n}\right)$, where $n$ is large.

## END-OF-CHAPTER REVIEW EXERCISE 5

1 The mean and standard deviation of the time spent by visitors at an art gallery are 3.5 hours and 1.5 hours, respectively.

a Find the probability that the mean time spent in the art gallery by a random sample of:

i 60 people is more than 4 hours **[3]**

ii 20 people is less than 4 hours. **[3]**

b What assumption(s), if any, did you need to make in part **a ii**? **[1]**

2 The score on a four-sided spinner is given by the random variable $X$ with probability distribution as shown in the table.

| $X$ | 2 | 3 | 4 | 5 |
|---|---|---|---|---|
| $P(X = x)$ | 0.1 | 0.4 | 0.2 | 0.3 |

a Show that the variance is 1.01. **[3]**

b The spinner is spun 100 times and each score noted. Let $S$ be the random variable for the sum of 100 observations. Write down the approximate distribution of $S$. **[2]**

c Use a normal distribution to work out the probability that the sum of the 100 observations is less than 350. Explain why you can use the normal distribution in this situation. **[4]**

3 The burn time, in minutes, for a certain brand of candle can be modelled by a normal distribution with mean 90 and standard deviation 15.6. Find the probability that a random sample of five candles, each one lit immediately after another burns out, will burn for a total of 500 minutes or less. **[5]**



4 A random sample of 35 observations is to be taken from a normal distribution with mean 15 and variance 9. If $\bar{X}$ is the sample mean, find:

a $P(\bar{X} < 16.2)$ **[4]**

b the value of $k$, where $P(\bar{X} < k) = 0.75$. **[4]**

5 There are 12 equally talented children at a sports club. Jamil wishes to choose one child at random from these children to represent the club. The children are numbered 1, 2, 3 and so on up to 12. Jamil then throws two ordinary fair dice, each numbered 1 to 6, and he finds the sum of the scores. He chooses the child whose number is the same as the sum of the scores.

a Explain why this is a biased method of choosing a child. **[2]**

b Describe briefly an unbiased method of choosing a child. **[2]**

6 Dominic wishes to choose a random sample of five students from the 150 students in his year. He numbers the students from 1 to 150. Then he uses his calculator to generate five random numbers between 0 and 1. He multiplies each random number by 150 and rounds up to the next whole number to give a student number.

i Dominic's first random number is 0.392. Find the student number that is produced by this random number. **[1]**

ii Dominic's second student number is 104. Find a possible random number that would produce this student number. **[1]**

iii Explain briefly why five random numbers may not be enough to produce a sample of five student numbers. **[1]**

*Cambridge International AS & A Level Mathematics 9709 Paper 73 Q2 November 2016*

**7** It is known that the number, $N$, of words contained in the leading article each day in a certain newspaper can be modelled by a normal distribution with mean 352 and variance 29. A researcher takes a random sample of 10 leading articles and finds the sample mean, $\bar{N}$, of $N$.

**i** State the distribution of $\bar{N}$, giving the values of any parameters. **[2]**

**ii** Find $P(\bar{N} > 354)$. **[3]**

*Cambridge International AS & A Level Mathematics 9709 Paper 73 Q1 November 2015*

**8** Jyothi wishes to choose a representative sample of 5 students from the 82 members of her school year.

**i** She considers going into the canteen and choosing a table with five students from her year sitting at it, and using these five people as her sample. Give two reasons why this method is unsatisfactory. **[2]**

**ii** Jyothi decides to use another method. She numbers all the students in her year from 1 to 82. Then she uses her calculator and generates the following random numbers.

231492 762305 346280

From these numbers, she obtains the student numbers 23, 14, 76, 5, 34 and 62. Explain how Jyothi obtained these student numbers from the list of random numbers. **[3]**

*Cambridge International AS & A Level Mathematics 9709 Paper 73 Q1 June 2015*

**9** The editor of a magazine wishes to obtain the views of a random sample of readers about the future of the magazine.

**i** A sub-editor proposes that they include in one issue of the magazine a questionnaire for readers to complete and return. Give two reasons why the readers who return the questionnaire would not form a random sample. **[2]**

The editor decides to use a table of random numbers to select a random sample of 50 readers from the 7302 regular readers. These regular readers are numbered from 1 to 7302. The first few random numbers which the editor obtains from the table are as follows.

49757 80239 52038 60882

**ii** Use these random numbers to select the first three members in the sample. **[2]**

*Cambridge International AS & A Level Mathematics 9709 Paper 73 Q2 November 2010*

**10** The lengths of time people take to complete a certain type of puzzle are normally distributed with mean 48.8 minutes and standard deviation 15.6 minutes. The random variable $X$ represents the time taken, in minutes, by a randomly chosen person to solve this type of puzzle. The times taken by random samples of 5 people are noted. The mean time $\bar{X}$ is calculated for each sample.

**i** State the distribution of $\bar{X}$, giving the values of any parameters. **[2]**

**ii** Find $P(\bar{X} < 50)$. **[3]**

*Cambridge International AS & A Level Mathematics 9709 Paper 7 Q2 June 2008*

# Chapter 6
# Estimation

**In this chapter you will learn how to:**

- calculate unbiased estimates of the population mean and variance from a sample
- formulate hypotheses and carry out a hypothesis test concerning the population mean in cases where the population is normally distributed with known variance or where a large sample is used
- determine and interpret a confidence interval for a population mean in cases where the population is normally distributed with known variance or where a large sample is used
- determine, from a large sample, an approximate confidence interval for a population proportion.

**PREREQUISITE KNOWLEDGE**

| Where it comes from | What you should be able to do | Check your skills |
|---|---|---|
| Probability & Statistics 1, Chapters 2 and 3 | Calculate the mean and variance from raw and summarised data. | Calculate the mean, variance and standard deviation for the following data sets:<br>**1** $n = 11$ $\sum x = 16.5$ $\sum x^2 = 25.85$<br>**2** $n = 8$ $\sum x = 434$ $\sum x^2 = 26\,630$<br>**3** 20 24 15 18 16 25 22<br>**4** 6.5 9.3 13.7 15.1 20.4 |
| Chapter 1 | Formulate and carry out hypothesis testing. | State the null and alternative hypotheses and test statistic for the following:<br>**5** $X \sim \text{N}(86, 16)$; sample value 84; two-tailed test at 10%<br>**6** $X \sim \text{N}(54, 3^2)$; sample value 50; one-tailed test at 5%<br>**7** $X \sim \text{N}(18, 3)$; sample value 20; one-tailed test at 1% |
| Probability & Statistics 1, Chapter 8 | Know how to approximate a binomial distribution by a normal distribution. | Express the following as approximate normal distributions:<br>**8** $X \sim \text{B}(42, 0.4)$<br>**9** $X \sim \text{B}(100, 0.55)$ |



## Why do we study estimation?

Chapter 5 explained that it is not always possible to collect data about every item in a population. There are many practical situations when it is necessary to use a sample to obtain information about a population. For example, an asthma attack may lead to a hospital admission. Sample data allow us to estimate the number of people likely to require a stay in hospital following an asthma attack. Studying the length of the hospital stay will allow us to estimate hospital staffing and other resources. In turn, this allows the hospital to assess its resources and plan for the needs of other patients.

Conservationists study only samples of the population of certain species to make predictions of their numbers. For example, the estimate of the population of mountain gorillas is that there are fewer than 800 left in the world. Snow leopards live in 12 countries in central Asia. Since the start of this century, the estimated number of snow leopards has decreased by about 20%. The actual numbers of snow leopards and mountain gorillas are unknown. These numbers are estimates.

Consider a study that claims two-thirds of adults living in a particular country are overweight. It is unlikely that every adult in that country was weighed; yet the study states they have evidence to justify their claim. That evidence comes from summary statistics from a sample of adults.

The summary statistics calculated from a sample, the sample mean and the sample variance, are used to draw conclusions about the whole population based on the evidence from the sample. These calculated summary statistics, since they only use part of a population, are estimates. To differentiate between sample statistics and population statistics, the following convention is used:

- Population parameters, such as mean and variance, use Greek letters $\mu$ and $\sigma^2$, respectively.
- Estimates of population parameters from a sample are written using Roman letters; for example, $\bar{x}$ is the sample mean and $s^2$ is the sample variance.

Note that the subject you are studying is 'statistics' and confusingly an estimate of a population parameter is called a **statistic**; so estimates of a population's mean and variance are called population statistics.

**REWIND**

Section 5.2 in the previous chapter explained about the sample mean, $\bar{x}$. This is an estimate for the mean, $\mu$, of a population. The sample mean is an unbiased estimate since the expected value of the sampling distribution of the sample mean is equal to the mean of the population, the parameter it is estimating.

As an example, suppose you wish to find out the average number of fiction books people read each month. You cannot ask the entire population, so instead you ask a sample of the population and work out the average number of fiction books read each month from the sample data. For an unbiased estimate you need to use an unbiased sampling method, such as random sampling, that ensures all members of the population have an equal chance of being selected for the sample; and, of course, you must ask unambiguous questions, making it clear that you are only interested in fiction and not non-fiction books.



## 6.1 Unbiased estimates of population mean and variance

One objective when taking a sample is to estimate population statistics. A statistic or estimate is a numerical value calculated from a set of data and used in place of an unknown parameter in a population. The bias of an estimate is the difference between the expected value of the estimate and the true value of the parameter. This difference is the **sampling error**. The most efficient estimate is one that is unbiased.

The reliability of an estimate can also depend on the variance of the population. A population with a small variance implies that the data are not widely dispersed and any sample is therefore less likely to be seriously unrepresentative. Conversely, a population with a large variance implies that the data are widely dispersed and so an unrepresentative sample may easily arise.

**KEY POINT 6.1**

A statistic is an estimate of a given population parameter, calculated from sample data.

A statistic is an unbiased estimate of a given population parameter when the mean of the sampling distribution of that statistic is equal to the parameter being estimated.

If $\hat{U}$ is some statistic derived from a random sample taken from a population, then $\hat{U}$ is an unbiased estimate for $U$ if $\mathrm{E}(\hat{U}) = U$.

The most efficient estimate is one that is unbiased and has the smallest variance.

All the examples presented in Chapter 5 involved a sample from a population with known variance. In practice, if you do not know the population mean, you are unlikely to know the true population variance either.

To explore the sampling distribution of the sample variance, we can return to the example about the spinner numbered 1, 1, 2, 4. In Section 5.2, we found that this distribution has mean 2 and variance 1.5.

To explore the variance as the statistic, for a sample size of 1 we can work out the expectation of the variance $E(V)$.

| **Sample outcomes** | $\Sigma x^2$ | $\bar{x}$ | **Variance,** $v = \frac{\Sigma x^2}{1} - \bar{x}^2$ | **Probability (outcome)** |
|---|---|---|---|---|
| 1 | 1 | 1 | 0 | $\frac{1}{2}$ |
| 2 | 4 | 2 | 0 | $\frac{1}{4}$ |
| 4 | 16 | 4 | 0 | $\frac{1}{4}$ |

The sample variance, for sample size of 1, $E(V) = 0$.

This is not equal to the variance of the original population, so the sample variance is not an unbiased estimate for the variance.

We do not need to explore the variance for another sample size since a single example that shows the variance is biased is sufficient to prove the point. However, it is worthwhile exploring other sample sizes to see if there is a possible connection between the sample variance and the population variance.

For a sample of size 2, first list all possible sample outcomes, together with the variance and probability of choosing that sample.

| **Sample outcomes** | $\Sigma x^2$ | $\bar{x}$ | **Variance,** $v = \frac{\Sigma x^2}{2} - \bar{x}^2$ | **Probability (outcome)** |
|---|---|---|---|---|
| 1 1 | 2 | 1 | 0 | $\frac{4}{16}$ |
| 2 2 | 8 | 2 | 0 | $\frac{1}{16}$ |
| 4 4 | 32 | 4 | 0 | $\frac{1}{16}$ |
| 1 2 | 5 | $1\frac{1}{2}$ | $\frac{1}{4}$ | $\frac{4}{16}$ |
| 1 4 | 17 | $2\frac{1}{2}$ | $2\frac{1}{4}$ | $\frac{4}{16}$ |
| 2 4 | 20 | 3 | 1 | $\frac{2}{16}$ |

You can check the values for the probabilities of these sample outcomes in Chapter 5, Section 5.2.

We can now draw a probability distribution table for the sample variance, sample size of 2.

| $v$ | 0 | $\frac{1}{4}$ | 1 | $2\frac{1}{4}$ |
|---|---|---|---|---|
| $P(V = v)$ | $\frac{3}{8}$ | $\frac{1}{4}$ | $\frac{1}{8}$ | $\frac{1}{4}$ |

Hence, $E(V) = \left(0 \times \frac{3}{8}\right) + \left(\frac{1}{4} \times \frac{1}{4}\right) + \left(1 \times \frac{1}{8}\right) + \left(2\frac{1}{4} \times \frac{1}{4}\right) = \frac{3}{4}$

Comparing this result with the variance of the original population, we find that $E(V) = \frac{3}{4}$ and $\sigma^2 = 1\frac{1}{2}$, the variance of the original population.

Notice that $\frac{3}{4} = \frac{1}{2} \times 1\frac{1}{2}$, or $E(V)\,\frac{n-1}{n} \times \sigma^2$, where $n$ is the sample size.

We need more than just this example to see if this relationship between the original variance and the estimate of variance always holds.

Here are the data for a sample size of 3. You can refer to Chapter 5, Section 5.2 for outcomes and probabilities.

| Sample outcome | $\Sigma x^2$ | $\bar{x}$ | Variance, $v = \frac{\Sigma x^2}{3} - \bar{x}^2$ | Probability (outcome) |
|---|---|---|---|---|
| 1 1 1 | 3 | 1 | 0 | $\frac{1}{8}$ |
| 1 1 2 | 6 | $\frac{4}{3}$ | $\frac{2}{9}$ | $\frac{3}{16}$ |
| 1 2 2 | 9 | $\frac{5}{3}$ | $\frac{2}{9}$ | $\frac{3}{32}$ |
| 2 2 2 | 12 | 2 | 0 | $\frac{1}{64}$ |
| 1 1 4 | 18 | 2 | 2 | $\frac{3}{16}$ |
| 1 2 4 | 21 | $\frac{7}{3}$ | $\frac{14}{9}$ | $\frac{6}{32}$ |
| 2 2 4 | 24 | $\frac{8}{3}$ | $\frac{8}{9}$ | $\frac{3}{64}$ |
| 1 4 4 | 33 | 3 | 2 | $\frac{3}{32}$ |
| 2 4 4 | 36 | $\frac{10}{3}$ | $\frac{8}{9}$ | $\frac{3}{64}$ |
| 4 4 4 | 48 | 4 | 0 | $\frac{1}{64}$ |

The probability distribution table for the sample variance, sample size of 3, is therefore:

| $v$ | 0 | $\frac{2}{9}$ | $\frac{8}{9}$ | $\frac{14}{9}$ | 2 |
|---|---|---|---|---|---|
| $P(V = v)$ | $\frac{5}{32}$ | $\frac{9}{32}$ | $\frac{3}{32}$ | $\frac{6}{32}$ | $\frac{9}{32}$ |

Hence, $E(V) = \left(0 \times \frac{5}{32}\right) + \left(\frac{2}{9} \times \frac{9}{32}\right) + \left(\frac{8}{9} \times \frac{3}{32}\right) + \left(\frac{14}{9} \times \frac{6}{32}\right) + \left(2 \times \frac{9}{32}\right) = 1$, and if we use the relationship $E(V) = \frac{n-1}{n} \times \sigma^2$ for sample size of 3, we find that $E(V) = \frac{3-1}{3} \times 1\frac{1}{2} = 1$, the same value.

In general, for a population where $\sigma^2$ is the population variance, the expectation of sample variance $E(V)$ is given by:

$$E(V) = \frac{n-1}{n} \times \sigma^2$$

Using the results we met in Chapter 3, Key point 3.3, this means that $E\left(\frac{nV}{n-1}\right) = \sigma^2$ and $\frac{nV}{n-1} = \frac{n}{n-1}\left(\frac{\Sigma X^2}{n} - \bar{X}^2\right) = \frac{1}{n-1}\left(\Sigma X^2 - n\bar{X}^2\right)$.

**KEY POINT 6.2**

For sample size $n$ taken from a population, an unbiased estimate of the population mean $\mu$ is the sample mean $\bar{x}$.

An unbiased estimate of the population variance $\sigma^2$ is:

$$s^2 = \frac{1}{n-1}\left(\Sigma x^2 - n\bar{x}^2\right)$$

**TIP**

Data may be raw data or summarised data. Use one of the equivalent formulae for variance to suit the information:

$$\frac{1}{n-1}\left(\Sigma x^2 - n\bar{x}^2\right)$$

$$= \frac{1}{n-1}\left(\Sigma x^2 - \frac{(\Sigma x)^2}{n}\right)$$

$$= \frac{1}{n-1}\left(\Sigma(x-\bar{x})^2\right)$$

To find an unbiased estimate of variance on your calculator, use $\sigma_{n-1}$ or $s_{n-1}$.



**WORKED EXAMPLE 6.1**

A conservationist wishes to estimate the variance of the numbers of eggs laid by Melodious larks. The following data summarise her results for a sample of 30 Melodious larks' nests ($m$).

$$\Sigma m^2 = 162, \ \Sigma m = 66$$

Use the data to find an unbiased estimate for the variance of the number of eggs laid by Melodious larks.

**Answer**

$$\frac{1}{n-1}\left(\Sigma m^2 - \frac{(\Sigma m)^2}{n}\right) = \frac{1}{30-1}\left(162 - \frac{66^2}{30}\right) = 0.579$$

The summarised data suggest using $\frac{1}{n-1}\left(\Sigma x^2 - \frac{(\Sigma x)^2}{n}\right)$.

Note that if the question had stated: 'The following data summarise the results of number of eggs laid in 30 nests of Melodious larks. $\Sigma m^2 = 162$, $\Sigma m = 66$. Find the variance.', then the population would be just that group of nests and you would use the variance formula

$$\frac{1}{n}\left(\Sigma x^2 - \frac{(\Sigma x)^2}{n}\right) = \frac{1}{30}(162 - 145.2) = 0.56.$$

Note that $\frac{29}{30} \times 0.579... = 0.56$

**WORKED EXAMPLE 6.2**

A team of conservationists monitoring a tiger population record the number of tiger cubs in a sample of 24 litters. The table shows their findings.

| Number of cubs, $c$ | 1 | 2 | 3 | 4 | $>4$ |
|---|---|---|---|---|---|
| Frequency, $f$ | 2 | 7 | 12 | 3 | 0 |

Find unbiased estimates for the mean and variance of the number of tiger cubs in the litters.

**Answer**

$$\bar{c} = \frac{\sum fc}{\sum f} = \frac{(1\times 2)+(2\times 7)+(3\times 12)+(4\times 3)}{24}$$

$$= \frac{64}{24} = 2\frac{2}{3}$$

Adapt formulae for grouped frequency.

Unbiased estimate for mean is equal to the sample mean.

You can input data into the calculator and use $\sigma_{n-1}$ or $s_{n-1}$. Show key values in your working.

$$s^2 = \frac{1}{n-1}\left(\sum fc^2 - \frac{\left(\sum fc\right)^2}{n}\right)$$

$$= \frac{1}{24-1}\left[\left((2\times 1^2)+(7\times 2^2)\right)+(12\times 3^2)+(3\times 4^2)-\frac{64^2}{24}\right]$$

$$= \frac{1}{23}\left[186-\frac{4096}{24}\right] = \frac{2}{3}$$



**EXERCISE 6A**

In all of the following questions you are given some data and some descriptive statistics of the data. Your task is to find unbiased estimates of the population mean and variance in each question.

1 Data: the length, $x$ cm, of an electrical component.

$n = 32$, $\sum x = 70.4$ and $\sum x^2 = 175.56$

2 Data: the time taken, $t$ minutes, in a random sample of dental check-up appointments.

$n = 30$, $\sum t = 630$ and $\sum t^2 = 13\,770$

3 Data: the yield per plant, in kg rounded to the nearest 100 g, of a random sample of a variety of aubergine plants.

3.5 3.7 4.1 4.4 4.6 4.5 4.5 4.3 4.2

4 Data: the volumes, in ml, for a brand of ice cream in a 750 ml container.

748 751 748 751 745 756 753 760

5 Data: the total mass, $x$ grams, for a random sample of quail eggs.

$n = 16$, $\sum x = 128.4$ and $\sum x^2 = 1137.6$

6 Data: the total number of faults found in a random sample of 60 silk scarves.

| Number of faults per silk scarf | 0 | 1 | 2 | 3 | 4 |
|---|---|---|---|---|---|
| Number of silk scarves | 26 | 16 | 7 | 8 | 3 |

7 Data: the time taken, in days, for a random sample of letters posted second class to be delivered.

| Number of days for letter to be delivered | 1 | 2 | 3 | 4 | 5 |
|---|---|---|---|---|---|
| Number of letters | 24 | 32 | 29 | 9 | 6 |

## 6.2 Hypothesis testing of the population mean

Sample data are often collected to test a statistical hypothesis about a population. Such a sample, even if it is a random sample, may or may not be representative of the population. The central limit theorem studied in Chapter 5 proves that random sample estimates can be used to make statements about populations without having to assume that the populations have normal distributions. Estimates of the sample mean and sample variance can be calculated from the sample and these estimates can be used to see if they support or reject the null hypothesis. For sample data, $\sqrt{(\text{sample variance})}$; that is, $\frac{\sigma}{\sqrt{n}}$, is referred to as the **standard error**.

We follow the same process as previously used when carrying out a hypothesis test of the population mean. Ideally, we will set up the hypotheses, then collect the sample of data in that order.

**REWIND**

The steps to carry out a hypothesis test are explained in Chapter 1, but in summary they are:

- Decide whether the situation calls for a one-tailed or two-tailed test.
- State the null and alternative hypotheses.
- Decide on the significance level.
- Calculate the test statistic.
- Compare the calculated probability with the critical value(s).
- Interpret the result in terms of the original claim.

You may also need to consider Type I and Type II errors:

- A Type I error occurs when a true null hypothesis is rejected.
- A Type II error occurs when a false null hypothesis is accepted.

### Hypothesis test of a population mean from a normal population with known variance

**KEY POINT 6.3**

If the population mean is unknown, but the population variance is known, sample data can be used to carry out a hypothesis test that the population mean has a particular value, as follows:

For a sample size $n$ drawn from a normal distribution with known variance, $\sigma^2$, and sample mean $\bar{x}$, the test statistic is:

$$z = \frac{\bar{x} - \mu}{\frac{\sigma}{\sqrt{n}}}$$

## WORKED EXAMPLE 6.3

The masses of cucumbers grown at a smallholding are normally distributed with mean 310 g and standard deviation 22 g. Producers of a new plant food claim that its use increases the masses of cucumbers. To test this claim, some cucumber plants are grown using the new plant food and a random sample of 40 cucumbers from these plants are selected and weighed. The mean mass of these cucumbers is 316 g.

Assuming the standard deviation of the masses of the sample is the same as the standard deviation of the population, test the claim at a 5% level of significance.

**Answer 1**

Let $\bar{X}$ be the mean mass of cucumbers.

First set up the test.

Then $\bar{X} \sim \mathrm{N}\left(310, \frac{22^2}{40}\right)$.

$\mathrm{H_0}$: $\mu = 310$

$\mathrm{H_1}$: $\mu > 310$

One-tailed test at 5% level of significance

Use a one-tailed test, as you are looking for an increase in weight.

$$\mathrm{P}(\bar{X} > 316) \approx \mathrm{P}\left(z > \frac{316 - 310}{\frac{22}{\sqrt{40}}}\right)$$

$$= 1 - \Phi(1.725)$$

$$= 0.0423 \text{ or } 4.23\%.$$

Calculate the test statistic using $z = \frac{\bar{x} - \mu}{\sigma/\sqrt{n}}$ and compare to significance level 5%.

4.23% < 5%, so the masses are in the critical region.
We reject $\mathrm{H_0}$, because there is some evidence to support the plant food producer's claim.

Comment on your result in the context of the question.

**Answer 2**

Alternative way of writing the solution:

An alternative approach is to compare the test statistic $z$ with the critical value from tables.

$\mathrm{H_0}$: $\mu = 310$

$\mathrm{H_1}$: $\mu > 310$

One-tailed test at 5% level of significance, critical value $z$ is 1.645.

Calculate the test statistic.

$$z = \frac{316 - 310}{\frac{22}{\sqrt{40}}} = 1.725 \text{ and } 1.725 > 1.645.$$

Compare it with the critical value, which is $z = \phi^{-1}(0.95) = 1.645$.

So reject $\mathrm{H_0}$ and accept $\mathrm{H_1}$.
There is some evidence to accept the plant food producer's claim.

Comment on your result in the context of the question.

## WORKED EXAMPLE 6.4

The burn time, in minutes, for a certain brand of candle is modelled by a normal distribution with standard deviation 5.7. The manufacturer claims that the mean is 250 minutes. Lanfen randomly selects ten of these candles and finds that their burn times in minutes are as follows:

245 247 236 255 250 239 241 252 251 243

Stating any assumptions you make, investigate at the 5% level of significance whether the manufacturer's claim is valid.

**Answer**

Assumptions:
Random sample chosen.
Standard deviation of the sample the same as the population.

The assumptions are the conditions that allow you to use a sample to investigate the claim.

$$\text{Mean} = \frac{1}{10}(245 + 247 + 236 + 255 + 250 + 239 + 241 + 252 + 251 + 243) = \frac{2459}{10} = 245.9$$

First find the mean of the sample.

$H_0$: $\mu = 250$

$H_1$: $\mu \neq 250$

State the null and alternative hypotheses.

Two-tailed test at 5% significance,
critical value $z$ is −1.96.

This is a two-tailed test, as Lanfen is not investigating whether the claim is only too high or only too low.

Use tables to find the critical value, which is $z = \Phi^{-1}(-0.975) = -1.96$.

$$\overline{X} \sim \mathrm{N}\left(250, \frac{5.7^2}{10}\right)$$

$$z = \frac{245.9 - 250}{\frac{5.7}{\sqrt{10}}} = -2.275$$

Calculate the test statistic.

$\Phi(-2.275) = 1 - 0.9886 = 0.0114 < 2.5\%$ or
$-2.275 < -1.96$

Compare the test statistic with the critical value.

Reject $H_0$. There is sufficient evidence to doubt the manufacturer's claim.

Comment on your result in the context of the question.

## Hypothesis test of population mean using a large sample



It is possible to carry out a hypothesis test of a population mean when the population variance is unknown. Provided the sample is large, we follow the same process as for a hypothesis test of a population mean from a normal population with known variance. For the variance, we use $s^2$, an unbiased estimate of the population variance, where

$$s^2 = \frac{1}{n-1}\left(\Sigma\, x^2 - \frac{(\Sigma\, x)^2}{n}\right).$$

**KEY POINT 6.4**

If the population mean and population variance are unknown, sample data can be used to conduct a hypothesis test that the population mean has a particular value, as follows:

For a large sample size $n$ drawn with unknown variance and sample mean $\bar{x}$, the test statistic is:

$$z = \frac{\bar{x} - \mu}{\frac{s}{\sqrt{n}}}$$

where $s^2 = \dfrac{1}{n-1}\left(\Sigma x^2 - \dfrac{(\Sigma x)^2}{n}\right)$.

## WORKED EXAMPLE 6.5

A researcher believes that students underestimate how long 1 minute is. To test his belief, 42 students are chosen at random. Each student, in turn, closes their eyes and estimates 1 minute. The results for their times, $x$ seconds, are summarised as follows:

$$\sum x = 2471 \text{ and } \sum x^2 = 146\,801$$

Investigate at the 10% level of significance if there is any evidence to support the researcher's claim. What advice would you give to the researcher based on your findings?

**Answer**

$H_0$: $\mu = 60$

$H_1$: $\mu < 60$

First, state the null and alternative hypotheses. Use 1 minute = 60 seconds.

One-tailed test at 10% level of significance, critical value $z$ is −1.282.

Decide whether a one-tailed or two-tailed test is appropriate and find the critical value using tables.

$$\bar{x} = \frac{\sum x}{n} = \frac{2471}{42} = 58.83$$

Find unbiased estimates for the mean and variance.

$$s^2 = \frac{1}{42-1}\left(146\,801 - \frac{2471^2}{42}\right) = 34.73$$

Use $\frac{1}{n-1}\left(\sum x^2 - \frac{(\sum x)^2}{n}\right)$ to find an unbiased estimate for the variance.

$$z = \frac{58.83 - 60}{\sqrt{\frac{34.73}{42}}} = -1.287$$

Next, calculate the test statistic.

$-1.287 < -1.282$

Compare the test statistic with the critical value and comment in context of the question.

Reject $H_0$ and accept $H_1$. There is evidence to support the researcher's claim.

The value of the test statistic and the critical value are very close; advise the researcher to do more tests.

Advise the researcher of your findings, always in context of the original problem.



## EXERCISE 6B

1 The manufacturer of a 'fast-acting pain relief tablet' claims that the time taken for its tablet to work follows a normal distribution with mean 18.4 minutes and variance $3.6^2$ minutes$^2$. Tyler claims that the tablets do not work that quickly. To test the claim, a random sample of 40 people record the time taken for the tablet to work. The mean time for this sample is 19.7 minutes.

Assuming the sample and population variances are the same, carry out an appropriate hypothesis test at the 1% level of significance.

2 IQ test scores are normally distributed and are designed to have a mean score of 100. Anna believes the mean is higher than 100. A random sample of 180 people's IQ test scores, $x$, are summarised as follows.

$\sum x = 18\,432$ and $\sum x^2 = 1\,926\,709.4$

a Carry out an appropriate hypothesis test at the 2% level of significance.

b Anna then discovers that the IQ test is also designed to have a variance of $15^2$. A random sample of six people take the test and their IQ test scores are:

103 109 112 96 100 104

To test Anna's belief, carry out an appropriate hypothesis test, using just the random sample of six people, at the 2% level of significance.

c Comment, with reasons, on the reliability of your answers to parts **a** and **b**.

3 The mass of pesto dispensed by a machine to fill a jar is a normally distributed random variable with mean 380 g. The variance of the mass, in grams$^2$, of the pesto in the jars is 6.4. Each week a check is made to see that the mean mass dispensed by the machine has not significantly reduced. One particular week a sample of ten jars is checked. The mean mass of pesto in these jars is 378.7 g. Carry out an appropriate hypothesis test at the 5% level of significance, stating any assumptions you have made.

4 The average mass of large eggs is 68 g. The variance of the masses, in grams$^2$, of large eggs is $1.7^2$. A farm shop sells large eggs singly. A customer claims that the eggs are underweight. To test the claim, a random sample of large eggs is weighed. Their masses, in grams, are as follows:

68 65 59 72 65 60 71 73

Carry out an appropriate hypothesis test at the 1% level of significance, stating any assumption(s) you have made.

5 A machine dispenses ice cream into a cone. The amount dispensed follows a normal distribution with mean 80 ml and the variance of the amount of ice cream dispensed, in ml$^2$, is 9. A consumer complains that the amount is too low. To check whether the machine is dispensing the correct amount, a sample of six cones is checked. The volumes in ml are as follows:

82 72 75 80 76 80

Carry out an appropriate hypothesis test at the 5% level of significance, stating any assumption(s) you have made.

6 A shop sells 2 kg bags of potatoes. A quality control inspection checks the masses of 80 randomly chosen bags. Their masses, $x$, are summarised as follows:

$\Sigma x = 158.14$ and $\Sigma x^2 = 314.094$

Assuming the masses of the bags of potatoes are normally distributed, investigate at the 5% level of significance whether there is any evidence that the bags are underweight.

7 A manufacturer claims its light bulbs last for an average of 2000 hours. A random sample of 42 light bulbs is tested. The lengths of time the light bulbs last, $t$ hours, are summarised as follows:

$\Sigma t = 83\,895$ and $\Sigma x^2 = 167\,589\,883.6$

Test the manufacturer's claim at the 10% level of significance, stating any assumptions you have made.

## 6.3 Confidence intervals for population mean

When a hypothesis test reveals statistically significant results, the results are applicable to the sample. Often we use the results as if they apply to the population. However, we cannot be certain that the sample is actually representative of the population.

The hypothesis tests studied so far, in Chapter 1, Chapter 2 and earlier in this chapter, involve a single parameter, the population mean, from a sample of data. To allow for the issue that the sample may or may not be representative of the whole population, sample data can also be used to construct an interval that specifies the limits within which it is likely that the population mean will lie. This interval is a **confidence interval (CI)**.
A confidence interval for a parameter is constructed at a $P$% level of confidence such that if the same population is sampled many times and each time an interval estimate is found, the true population parameter will occur in $P$% of those intervals.

It is possible to construct one-sided or two-sided confidence intervals. However, we will consider only symmetrical two-sided intervals.

A 95% confidence interval is the range of values in which we can be 95% confident that the true mean lies. If that interval is from $a$ to $b$, then: P($a <$ true mean $< b$) = 0.95. The central 95% of the sample distribution is from the 2.5th to the 97.5th percentile.

For a normal distribution $N(\mu, \sigma^2)$, we find from normal tables that the central 95% lies between −1.96 and +1.96 standard deviations either side of the mean.

For a sample distribution, we use $N\left(\mu, \frac{\sigma^2}{n}\right)$.

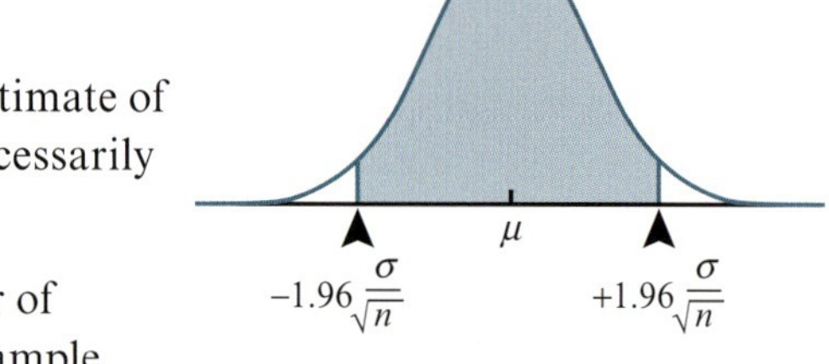


Although the sample mean, $\bar{x}$, is an unbiased estimate of the true mean, it is only an estimate. It is not necessarily the true mean.

If we work out sample means for a large number of samples, 95% of the time we would expect the sample mean, $\bar{x}$, to lie within the shaded area; that is:

$$\mu - 1.96\frac{\sigma}{\sqrt{n}} < \bar{x} < \mu + 1.96\frac{\sigma}{\sqrt{n}}$$

which rearranges to give:

$$\bar{x} - 1.96\frac{\sigma}{\sqrt{n}} < \mu < \bar{x} + 1.96\frac{\sigma}{\sqrt{n}}$$



So to find a 95% confidence interval, use the sample values and work out the interval $\bar{x} \pm 1.96\frac{\sigma}{\sqrt{n}}$. An alternative way to write the interval is $\left(\bar{x} - 1.96\frac{\sigma}{\sqrt{n}}, \bar{x} + 1.96\frac{\sigma}{\sqrt{n}}\right)$.

**KEY POINT 6.5**

A 95% confidence interval means that 95% of possible sample means lie within the interval. It tells us the probability that the true mean lies within the interval is 0.95, and the probability that the true mean does not lie within the interval is $1 - 0.95 = 0.05$.

## Confidence intervals for a population mean from a normal population with known variance

Consider Worked example 6.3. The hypothesis test found that there was evidence to accept the producer's claim that its plant food increases the mass of cucumbers.

The sample data are summarised by sample mean $\bar{x} = 316$ and standard error $\frac{22}{\sqrt{40}} = 3.48$.

A 95% confidence interval for these values is $316 \pm 1.96 \times 3.48$; that is, (309, 323).

This means we can be 95% confident that the true mean lies in this range.

Before using the new plant food, the mean mass of cucumbers was 310 g. This mass just lies within the confidence interval (309, 323), at the lower end. We could conclude that it is possible that the plant food does not increase the mass of the cucumbers and the sample is not representative.

The percentage level chosen for the confidence interval does affect the size of the interval. For example, consider what happens with a 90% confidence interval.

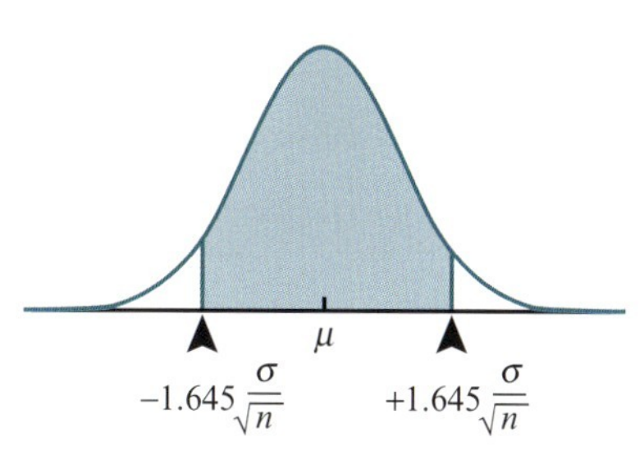


A 90% confidence interval will give a smaller interval.

From normal tables, the central 90% lies within 1.645 standard deviations of the mean.

The 90% confidence interval is $316 \pm 1.645 \times 3.48$; that is, (310, 322).

The lower bound 310 has been rounded from 310.3. Using a 90% confidence interval, the original population mean 310 lies just outside the interval and so you would accept the producer's claim.

What about a 99% confidence interval? Using normal tables, the 99% confidence interval can be calculated as $\mu \pm 2.576\frac{\sigma}{\sqrt{n}}$, giving $316 \pm 2.576 \times 3.48$; that is, (307, 325).

Compare the confidence intervals, with all values given to the nearest integer: 90% CI = (310, 322), 95% CI = (309, 323) and 99% CI = (307, 325). We can see that the higher the percentage, the more confident we can be that the true mean lies within that interval. However, the higher percentage gives a wider interval, and this means the information we have about the true mean is less precise; that is, there is a greater range of possible values for the true mean.

Sample size also affects the size of a confidence interval.

Consider a population with known standard deviation 15. A random sample $n = 100$ and $\bar{x} = 20$ has standard error $\frac{\sigma}{\sqrt{n}} = \frac{15}{\sqrt{100}} = 1.5$. A 95% confidence interval is then $20 \pm 1.96 \times 1.5$ or $(17.1,\ 22.9)$.

Let us increase $n$. If $n = 400$ with $\bar{x} = 20$, then standard error $\frac{\sigma}{\sqrt{n}} = \frac{15}{\sqrt{400}} = 0.75$ and the 95% confidence interval is narrower, $20 \pm 1.96 \times 0.75$ or $(18.5, 21.5)$.



## KEY POINT 6.6

A confidence interval for an unknown population parameter, such as the mean, at a $P$% confidence level, is an interval constructed so that there is a probability of $P$% that the interval includes the parameter.

To find the confidence interval for a population mean with known variance $\sigma^2$, calculate $\bar{x} \pm k\frac{\sigma}{\sqrt{n}}$, where $k$ is determined by the percentage level of the confidence interval.

| % CI | 90 | 95 | 98 | 99 |
|---|---|---|---|---|
| $k$ | 1.645 | 1.960 | 2.326 | 2.576 |

The greater the percentage, the more confident we can be that the true parameter lies within the interval.

The greater the percentage, the wider the confidence interval and the less precise we can be about the value of the true parameter.

When choosing the sample size, $n$, as $n$ increases the standard error $\frac{\sigma}{\sqrt{n}}$ decreases and the resulting confidence interval becomes narrower.

**EXPLORE 6.1**

As sample size increases, the value of $\frac{\sigma}{\sqrt{n}}$ decreases and so the width of a confidence interval decreases. Why do you think it is not usual practice to use very large samples? Hint: Find the proportional decrease in the width of the confidence interval for different values of $n$.

Discuss these two questions in relation to the scenarios that follow:

- How large a sample do you actually need?
- What confidence level are you prepared to accept?

Scenario 1: Health officials for a city with population around 40 000 are concerned with the increase in body mass index, BMI, in the population. Would your sample numbers change if the population was, say, 120 000? Explain why or why not.

Scenario 2: A health body wishes to investigate the effectiveness of a new drug treatment. Discuss the possible advantages and disadvantages in combining several different trials of a new drug treatment. (Note that combining the results of many scientific studies is called a meta-analysis.)

**WORKED EXAMPLE 6.6**

Excessive vegetation in pond water can cause the appearance of unwanted organisms. Over a long period of time it has been found that the number of unwanted organisms in 100 ml of pond water is approximately normally distributed with standard deviation 12. Adam takes six random 100 ml samples of water from his pond. The numbers of unwanted organisms in the samples are 56, 102, 48, 74, 88 and 67.

**a** Find a 95% confidence interval for the mean number of organisms in 100 ml of the pond water.

**b** If the mean number of unwanted organisms in 100 ml of pond water is above 80, vegetation should be removed. Use your results to decide whether Adam needs to remove vegetation from his pond. What advice would you give Adam?

**Answer**

**a** $\bar{x} = \frac{1}{6}(56 + 102 + 48 + 74 + 88 + 67) = 72.5$

First, find the mean of the sample.

$n = 6, \frac{\sigma}{\sqrt{n}} = \frac{12}{\sqrt{6}} = 4.9$

Then find the standard error.

$\text{CI} = 72.5 \pm (1.96 \times 4.9)$ or $(62.9, 82.1)$

Use $\bar{x} \pm 1.96\frac{\sigma}{\sqrt{n}}$ for a 95% confidence interval.

**b** The upper value of the range of values likely to contain the mean is 82.1. The probability that the mean of 80 organisms lies within this interval is 0.95. Despite the sample mean 72.5 being less than 80, the true mean could be as high as 82.1. Advise Adam to remove some of the vegetation.

Use your result to comment in context, to explain and justify the advice.

Advise Adam to take more samples of pond water.

**WORKED EXAMPLE 6.7**

The label on a certain packet of sweets states the contents are 100 g. It is known that the standard deviation is 5 g. The mechanism producing these packets of sweets is checked. From a random sample of ten packets, the mean is 103.8 g. Find a 99% confidence interval for the mean contents of the packets of sweets. Use your result to explain whether the mechanism needs adjustment.

**Answer**

$\bar{x} \pm 2.576 \frac{\sigma}{\sqrt{n}} = 103.8 \pm 2.576 \frac{5}{\sqrt{10}} = 103.8 \pm 4.073$ — Use $\bar{x} \pm 2.576 \frac{\sigma}{\sqrt{n}}$.

The CI is $(99.7,\ 107.9)$.

The confidence interval tells us that it is possible for the true mean to be below 100 g. The mechanism may need adjustment. — Comment in context. It is important that consumers do not get less than advertised.

**DID YOU KNOW?**

Quality control of manufacturing processes is one application of sampling methods. Random samples of the output of a manufacturing process are statistically checked to ensure the product falls within specified limits and consumers of the product get what they pay for. With any product, there can be slight variations in some parameter, such as in the radius of a wheel bolt. Statistical calculations using the distribution of frequent samples, usually chosen automatically, will give information to suggest whether the manufacturing process is working correctly.



## Confidence intervals for a population mean using a large sample

To find a confidence interval for a population mean, we rely upon knowing the standard deviation, $\sigma$, of the original population. However, since calculating standard deviation involves knowing the mean, it is more likely that the actual value of the standard deviation will be unknown. Instead, we can use the sample data to calculate an unbiased estimate of variance, $s^2$, and then use $s$ in place of $\sigma$ to find the confidence interval.

The procedure for finding a confidence interval using an unbiased estimate of standard deviation from a sample gives a reasonably accurate result provided the sample is sufficiently large. How large is sufficiently large? Look back at Explore 6.1. If you compare sample sizes 25 and 400, then since $\sqrt{25} = 5$ and $\sqrt{400} = 20$ you will find that increasing the sample size by 16 times $(16 \times 25 = 400)$ only reduces the margin of error by one-quarter $\left(\frac{1}{\sqrt{25}} \times \frac{1}{4} = \frac{1}{\sqrt{400}}\right)$. 'Large' is not precisely defined; as a general rule, it can be taken to be a sample size of 30 or more.

**KEY POINT 6.7**

To find the confidence interval for a population mean using a large sample, calculate $\bar{x} \pm k\frac{s}{\sqrt{n}}$, where $s = \sqrt{\frac{1}{n-1}\left(\sum x^2 - n\bar{x}^2\right)}$ and $k$ is determined by the percentage level of the confidence interval.

**WORKED EXAMPLE 6.8**

A sample of 60 strawberries is weighed, in grams. The results are summarised as follows:

$$\sum x = 972 \quad \text{and} \quad \sum x^2 = 17\,304.78$$

**a** Find a 90% confidence interval for the mean mass of the strawberries.

**b** An $\alpha$% confidence interval for the population mean, based on this sample, is found to have width of 3.65 grams. Find $\alpha$.

**Answer**

**a** $\bar{x} = \frac{\sum x}{n} = \frac{972}{60} = 16.2$

Find unbiased estimates for the mean and variance.

$s^2 = \frac{1}{60-1}\left(17\,304.78 - \frac{972^2}{60}\right) = 26.413$, so

Use $\frac{1}{n-1}\left(\sum x^2 - \frac{\left(\sum x\right)^2}{n}\right)$ to find an unbiased estimate for the variance.

$s = \sqrt{26.413} = 5.14$

$\bar{x} \pm 1.645\frac{s}{\sqrt{n}} = 16.2 \pm 1.645\frac{5.14}{\sqrt{60}} = 16.2 \pm 1.09$

The sample is sufficiently large to use $\bar{x} \pm 1.645\frac{s}{\sqrt{n}}$.

The CI is $(15.1,\ 17.3)$.

The confidence interval will be approximate, as the population standard deviation is unknown.

**b** $2k\frac{5.14}{\sqrt{60}} = 3.65$

The width of the CI is $\pm k\frac{s}{\sqrt{n}}$ or $2k\frac{s}{\sqrt{n}}$. Here you can use the value of $s$ from part **a**.

$k = \frac{3.65 \times \sqrt{60}}{2 \times 5.14} = 2.75$

For $z = 2.75$, from tables $p = 0.997$.

so $\alpha = 99.4\%$.

$1 - p$ gives the percentage in one tail.

$2(1-p)$ is the percentage in both tails.

$\alpha = 1 - 2(1-p)$

**EXERCISE 6C**

For questions 1 and 2 you may refer to the answers to Exercise 6A for unbiased estimates of population mean and variance. Give all confidence limits correct to 3 significant figures.

1 The following data summarise the length, $x$ cm, of an electrical component:

$n = 32$, $\sum x = 70.4$ and $\sum x^2 = 175.56$

Calculate:

**a** a 98% confidence interval for the population mean

**b** a 90% confidence interval for the population mean.

**2** The following data summarise the time taken, $t$ minutes, in a random sample of dental check-up appointments:

$n = 30$, $\Sigma t = 630$ and $\Sigma t^2 = 13\,770$

Calculate:

**a** a 95% confidence interval for the population mean

**b** a 98% confidence interval for the population mean.

**3** The following data summarise the total mass, $x$ grams of the yield for a random sample of 44 chilli plants.

$\Sigma x = 842$ and $\Sigma x^2 = 16\,364$

Calculate:

**a** a 99% confidence interval for the population mean

**b** a 95% confidence interval for the population mean.

**4** The following data summarise the volume, $x$ litres, for a random sample of bottles of juice.

$n = 68$, $\Sigma x = 134.14$ and $\Sigma x^2 = 266.094$

Calculate:

**a** a 90% confidence interval for the population mean

**b** a 99% confidence interval for the population mean.



**5** The following data summarise the total mass, $x$ grams, for a random sample of quail eggs:

$n = 30$, $\Sigma x = 254.4$ and $\Sigma x^2 = 2271.6$

**a** Calculate a 99% confidence interval for the population mean.

**b** An $\alpha$% confidence interval for the population mean, based on this sample, is found to have width of 1.3 grams. Find $\alpha$.

**6** The following data summarise the masses, $x$ kg, of 60 bags of dry pet food.

$\Sigma x = 117$ and $\Sigma x^2 = 232.72$

**a** Calculate unbiased estimates for the population mean and variance.

**b** Calculate a 98% confidence interval for the population mean.

**c** An $\alpha$% confidence interval for the population mean, based on this sample, is found to have width of 0.118 kg. Find $\alpha$.

M **7** **a** Explain why the width of a 98% confidence interval for the mean of a standard normal distribution is 4.652.

**b** The result, $X$, of testing the breaking strain of a brand of fishing line is a normally distributed random variable with mean $\mu$ and variance 2.25. The testers wish to have a 98% confidence interval for $\mu$ with a total width less than 1. Find the least number of tests needed.

## 6.4 Confidence intervals for population proportion

Not every statistical investigation concerns means of samples. Consider, for example, opinion polls. Many organisations carry out opinion polls to gauge voter intentions. Different polls for the same election do not always agree, even when the people chosen are representative samples of the population. When the data have been collected and presented, it is possible to calculate probabilities. We need to question, and statistically calculate, the reliability of their results.

There are a number of situations in which people are required to choose between two options, such as the UK Brexit vote where the options were to either remain or leave the European Union. The following example models an opinion poll for such a situation.

Sofia and Diego are the only two candidates in an election; there is no third option and *everyone* has to vote. Let a vote for Sofia be called a success. In an opinion poll of $n$ people, where $r$ people say they will vote for Sofia, the proportion of successes $\hat{p} = \frac{r}{n}$.

The binomial distribution is a suitable model for this situation since there are only two outcomes, there is a fixed number of people in the poll and each person independently chooses who to vote for.

Let the random variable, $X$, be the number of people who vote for Sofia. Then $X \sim \text{B}(n, p)$, $\text{E}(X) = np$ and $\text{Var}(X) = np(1 - p)$.

Let $\hat{P}$ be the random variable 'the proportion of the sample voting for Sofia'. Then $\hat{P} = \frac{X}{n}$.



Expected value, $\text{E}\left(\hat{P}\right) = \text{E}\left(\frac{X}{n}\right) = \frac{1}{n}\text{E}(X) = \frac{1}{n} \times np = p$, and so $\hat{p}$ is an unbiased estimate for $p$.

Variance, $\text{Var}\left(\hat{P}\right) = \text{Var}\left(\frac{X}{n}\right) = \frac{1}{n^2}\text{Var}(X) = \frac{1}{n^2} \times np(1 - p) = \frac{p(1-p)}{n}$

Before the election, an opinion poll of a random sample of 200 people is conducted. In this opinion poll 108 people say they will vote for Sofia and 92 say they will vote for Diego. With more than half of the people in the sample voting for Sofia, you may conclude that Sofia will win the election. To investigate how reliable this conclusion is we would have to find a confidence interval for the population proportion.

For sufficiently large values of $n$, such that $np > 5$ and $n(1 - p) > 5$, a binomial distribution can be approximated by a normal distribution. So an approximate distribution of the sample proportion is $\text{N}\left(p, \frac{p(1-p)}{n}\right)$.

Confidence intervals for a population proportion are worked out in a similar way to those for the sample mean. We calculate $\hat{p} \pm k\sqrt{\frac{\hat{p}(1-\hat{p})}{n}}$, where $k$ is determined by the percentage level of the confidence interval.

Note that this is an approximate confidence interval since a population proportion has a binomial distribution, which is discrete, whereas the normal distribution is continuous. However, it is not necessary to apply continuity corrections when finding these confidence intervals.

Returning to the opinion poll for Sofia and Diego, for the random sample of 200 people:

Sample proportion, $\hat{p} = \frac{108}{200} = 0.54$

Sample variance $= \frac{\hat{p}(1-\hat{p})}{n} = \frac{0.54 \times (1 - 0.54)}{200} = 0.001242$

For a 95% confidence interval:

$$\hat{p} \pm k\sqrt{\frac{\hat{p}(1-\hat{p})}{n}} = 0.54 \pm 1.96 \times \sqrt{0.001242} = 0.54 \pm 0.07$$

So the confidence interval is $(0.471, 0.609)$.

With only two candidates, the winner needs more than 50% of the votes. The question to be resolved is where the confidence interval lies with respect to the 50% value.

The following diagram shows the range of the confidence interval crossing the 50%, or 0.5, mark.

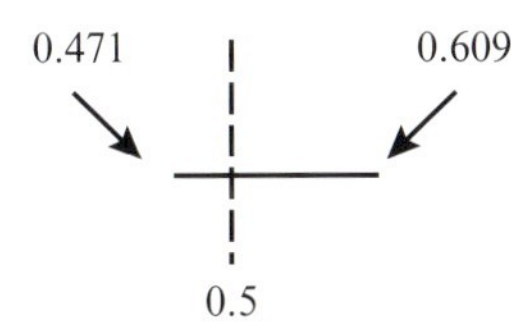


So for this sample, even though more than half said they would vote for Sofia, the confidence interval suggests that the proportion of votes for Sofia could be less than half.

Suppose instead you want to know how many people to poll (i.e. to select for the sample) to find a confidence interval of a given width. You could ask, 'What sample size is needed for an approximate 95% confidence interval for this proportion to have a width of 0.03?'.

To find a confidence interval, we calculate $\pm k\sqrt{\frac{\hat{p}(1-\hat{p})}{n}}$, so the width of the confidence interval is given by $2k\sqrt{\frac{\hat{p}(1-\hat{p})}{n}}$. The question requires the same proportion as the sample, so $\hat{p} = 0.54$.

For a 95% confidence interval we use $k = 1.96$.

$$2 \times 1.96\sqrt{\frac{0.54 \times (1-0.54)}{n}} = 0.03$$

$n = 4240$, to 3 significant figures.

**DID YOU KNOW?**

A claim by general election opinion polls is that they have a 3% margin of error. In practice, many such polls will have a sample size of 1000.

George Gallup showed the importance of opinion polls when he successfully predicted that Franklin Roosevelt would win the 1936 US presidential election. George Gallup continued to work in the field of public opinion. The work he began studying social, moral and religious opinions continues to this day in over 160 countries.

Today, opinion polls provide research and advice to many large organisations, such as manufacturers of new products, and political parties.

**WEB LINK**

You can find out more about the guidance of a large UK corporation on conducting and reporting of opinion polls on the BBC website.

**EXPLORE 6.2**

Find media reports on the results of an opinion poll. Does the report comment on how many people or voters were included in the poll? Does the report comment on the sampling method employed? Use the information given in the report to discuss the reliability of the results in the opinion poll.

**KEY POINT 6.8**

For a large random sample, size $n$, an approximate confidence interval for the population proportion, $\hat{p}$, is:

$$\left(\hat{p} - k\sqrt{\frac{\hat{p}(1-\hat{p})}{n}},\ \hat{p} + k\sqrt{\frac{\hat{p}(1-\hat{p})}{n}}\right)$$

where $k$ is determined by the percentage level of the confidence interval.

**WORKED EXAMPLE 6.9**

A Sudoku puzzle is classified as 'easy' if more than 70% of the people attempting to solve it do so within 10 minutes, and 'hard' if less than 20% of people take less than 10 minutes to complete it. Otherwise it is classified as 'average'. Of 120 people given a Sudoku puzzle, 87 completed it within 10 minutes.

**a** Find an approximate 99% confidence interval for the proportion of people completing the puzzle within 10 minutes. Comment on how the Sudoku puzzle should be classified.

**b** 200 random samples of 120 people were taken and a 99% confidence interval for the proportion was found from each sample. How many of these 200 confidence intervals would be expected to include the true proportion?

**Answer**

**a** $\hat{p} = \frac{87}{120} = 0.725$ — First, find the sample proportion.

$0.725 \pm 2.58\sqrt{\frac{0.725 \times (1 - 0.725)}{120}}$ — Use 2.58 for a 99% confidence interval.

$= 0.725 \pm 0.105$ — Calculate $\hat{p} \pm k\sqrt{\frac{\hat{p}(1-\hat{p})}{n}}$.

$\text{CI} = (0.62,\ 0.83)$ — Look where the confidence interval lies with respect to the boundaries of interest.

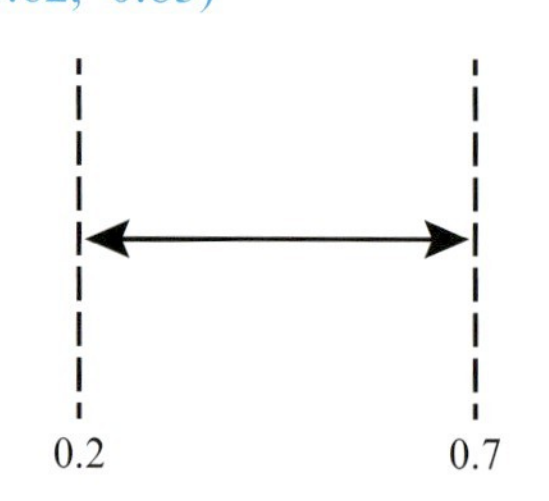


The CI crosses the 70% boundary, so although in the sample more than 70% of people completed it in less than 10 minutes, the interval suggests there will be some samples where less than 70% complete it in this time. It could be classified as easy or average.

**b** 198 — $200 \times 0.99$

**TIP**

Samples have a margin of error. When you find a confidence interval, you need to consider where it lies in relation to the boundary, or boundaries, used as guides for action.

## WORKED EXAMPLE 6.10

Apprentices work four days a week and spend one day a week at college. It is proposed that the college day is changed from a Monday to a Friday. The college will consider changing the day if 80% of apprentices are in favour of the change. In a sample of apprentices, how many should be asked to be 90% certain of gaining 80% support that is not more than 5% wrong.

**Answer**

$\hat{p} = 0.8$

You want the proportion in favour to be 80% or 0.8.

5% = 0.05

$1.645\sqrt{\frac{0.8 \times (1 - 0.8)}{n}} = 0.05$

Calculate $k\sqrt{\frac{\hat{p}(1-\hat{p})}{n}} = 0.05$.

Use 1.645 for 90% confidence interval.

$n = \left(\frac{1.645}{0.05}\right)^2 \times 0.8 \times (1 - 0.8) = 173$

## EXERCISE 6D

1 A quality control check of a random sample of 120 pairs of jeans produced at a factory finds that 24 pairs are sub-standard. Calculate the following confidence intervals for the proportion of jeans produced that are sub-standard:

   a a 90% confidence interval

   b a 98% confidence interval.

2 At a university, a random sample of 250 students is asked if they use a certain social media app. Of the students in the sample, 92 use this social media app. Calculate a 95% confidence interval for the proportion of students at the university who use this social media app.

3 A four-sided spinner has sides coloured red, yellow, green and blue. The probability that the spinner lands on yellow is $p$. In an experiment, the spinner lands on yellow 18 times out of 80 spins. Find an approximate 99% confidence interval for the value of $p$.

PS 4 A biased coin flipped 500 times results in tails 272 times.

   a Find a 90% confidence interval for the probability of obtaining a tail.

   b This experiment is carried out ten times. How many of the confidence intervals would be expected to contain the population proportion of obtaining a tail?

M 5 The proportion of European men who are red-green colour-blind is 8%. How large a sample would need to be selected to be 95% certain that it contains at least this proportion of red-green colour-blind men?

M PS 6 A random sample of 200 bees from a colony is tested to find out how many are infected with *Varroa* mites. Forty bees are found to be infected.

   a Calculate a 99% confidence interval for the proportion of the colony infected with *Varroa* mites.

   b The colony of bees will collapse and will not survive if 35% or more are infected with *Varroa* mites. Show why it is possible, at the 99% confidence level, that the colony of bees might collapse.

## Checklist of learning and understanding

- If $U$ is some statistic derived from a random sample taken from a population, then $U$ is an unbiased estimate for $\Phi$ if $E(U) = \Phi$.
- For sample size $n$ taken from a population, an unbiased estimate of:
  - the population mean $\mu$ is the sample mean $\overline{x}$
  - the population variance $\sigma^2$ is:

$$s^2 = \frac{1}{n-1}\left(\Sigma x^2 - n\overline{x}^2\right) = \frac{1}{n-1}\left(\Sigma x^2 - \frac{(\Sigma x)^2}{n}\right) = \frac{1}{n-1}\left(\Sigma(x-\overline{x})^2\right)$$

- To test a hypothesis about a sample mean, $\overline{x}$, for a sample size $n$ drawn from a normal distribution with known variance, $\sigma^2$, calculate the test statistic

$$z = \frac{\overline{x}-\mu}{\frac{\sigma}{\sqrt{n}}}.$$

  - The test statistic, $z$, can be used to test a hypothesis about a population mean drawn from any population.
  - Where the population variance is unknown, use the unbiased estimate of variance $s^2$.
- A confidence interval for an unknown population parameter, such as the mean, is an interval constructed so that it has a given probability that it includes the parameter.
- Confidence interval for:
  - population mean with known variance, $\sigma$, is $\left(\overline{x} - k\frac{\sigma}{\sqrt{n}},\ \overline{x} + k\frac{\sigma}{\sqrt{n}}\right)$
  - population mean using a large sample is $\left(\overline{x} - k\frac{s}{\sqrt{n}},\ \overline{x} + k\frac{s}{\sqrt{n}}\right)$, where $s = \sqrt{\frac{1}{n-1}\left(\Sigma x^2 - n\overline{x}^2\right)}$
  - population proportion, $\hat{p}$, is $\left(\hat{p} - k\sqrt{\frac{\hat{p}(1-\hat{p})}{n}},\ \hat{p} + k\sqrt{\frac{\hat{p}(1-\hat{p})}{n}}\right)$,

    where $k$ is determined by the percentage level of the confidence interval.

## END-OF-CHAPTER REVIEW EXERCISE 6

PS **1** The worldwide proportion of left-handed people is 10%.

**a** Find a 95% confidence interval for the proportion of left-handed people in a random sample of 200 people from town A. **[3]**

**b** In town B, there is a greater proportion of left-handed people than there is in town A. From a random sample of 100 people in town B, an $\alpha\%$ confidence interval for the proportion, $p$, of left-handed people is calculated to be (0.113, 0.207).

i Show that the proportion of left-handed people in the sample from town B is 16%. **[2]**

ii Calculate the value of $\alpha$. **[3]**

PS **2** The label on a jar of jam carries the words 'minimum contents 272 g'.

**a** Explain why, in practice, the average contents need to be greater than 272 g. **[1]**

**b** The mass of jam dispensed by a machine used to fill the jars is a normally distributed random variable with mean 276 g. The variance of the mass of the jam, in grams$^2$, dispensed by the machine is $1.8^2$. Each week there is a check to see if the mean mass dispensed by the machine is 276 g. One particular week a sample of eight jars is checked. The mean mass of jam in these jars is 277.7 g. Carry out an appropriate hypothesis test at the 5% level of significance, stating any assumptions you have made. **[5]**

M **3** An employer who is being sued for the wrongful dismissal of an employee is advised that any award paid out will be based on national average earnings for employees of a similar age. A random sample of 120 people is found to have a mean income of \$21 000 with standard error \$710.

**a** Find a 95% confidence interval for the award. **[3]**

**b** The employer wants to know the upper limit of the award that is very unlikely to be exceeded. The employer defines 'unlikely' as a probability of 0.001.

i Explain why the required size of the confidence interval is 99.8%. **[1]**

ii Work out the unlikely upper limit of the award, giving your answer to the nearest dollar. **[3]**



M **4** The volume, $v$ ml, of liquid dispensed by a vending machine for a random sample of 60 hot drinks is summarised as follows:

$$\Sigma v = 17\,280 \text{ and } \Sigma v^2 = 5\,015\,000$$

**a** Find unbiased estimates of the population mean and variance. **[2]**

**b** Work out a 90% confidence interval for the population mean. **[3]**

M PS **5** The manufacturer of a certain smartphone advertises that the average charging time for the battery is 80 minutes with standard deviation 2.6 minutes. Owners of these smartphones suggest that the time is longer. A random sample of the phones were charged from 0 to 100% and their times, in minutes, are as follows.

88 85 82 77 86 75 80 79

**a** Investigate at the 5% level of significance whether or not the manufacturer's claim is justified, stating any assumption(s) you have made. **[6]**

**b** The given length of time a charged smartphone battery will last is normally distributed with mean 24 hours. The variance of the time taken, in minutes, for the smartphone to work is 1. The variance of the length of time time taken, in hours squared, for the battery to last is 1. Sami tests a random sample of five batteries;

the sample mean time is 23.2 hours. Investigate at the 5% level of significance whether the time batteries last is less than the time given, stating any assumption(s) you make. **[5]**

**c** In a single sample, determine how long the battery could last for, if a Type I error occurs. **[1]**

**6** The manufacturer of a tablet computer claims that the mean battery life is 11 hours. A consumer organisation wished to test whether the mean is actually greater than 11 hours. They invited a random sample of members to report the battery life of their tablets. They then calculated the sample mean. Unfortunately a fire destroyed the records of this test except for the following partial document.

| Test of the mean batter the tablet | |
|---|---|
| Sample size, $n$ | |
| Sample mean (hours) | 11.8 |
| Is the result significant at the 5% level? | Yes |
| Is the result significant at the 2.5% level? | No |

Given that the population of battery lives is normally distributed with standard deviation 1.6 hours, find the set of possible values of the sample size, $n$. **[5]**

*Cambridge International AS & A Level Mathematics 9709 Paper 73 Q4 November 2016*

**7** Parcels arriving at a certain office have weights $W$ kg, where the random variable $W$ has mean $\mu$ and standard deviation 0.2. The value of $\mu$ used to be 2.60, but there is a suspicion that this may no longer be true. In order to test at the 5% significance level whether the value of $\mu$ has increased, a random sample of 75 parcels is chosen. You may assume that the standard deviation of $W$ is unchanged.

**i** The mean weight of the 75 parcels is found to be 2.64 kg. Carry out the test. **[4]**

**ii** Later another test of the same hypotheses at the 5% significance level, with another random sample of 75 parcels, is carried out. Given that the value of $\mu$ is now 2.68, calculate the probability of a Type II error. **[5]**

*Cambridge International AS & A Level Mathematics 9709 Paper 73 Q6 November 2015*

**8** Last year Samir found that the time for his journey to work had mean 45.7 minutes and standard deviation 3.2 minutes. Samir wishes to test whether his average journey time has increased this year. He notes the times, in minutes, for a random sample of 8 journeys this year with the following results.

46.2 41.7 49.2 47.1 47.2 48.4 53.7 45.5

It may be assumed that the population of this year's journey times is normally distributed with standard deviation 3.2 minutes.

**i** State, with a reason, whether Samir should use a one-tail or a two-tail test. **[2]**

**ii** Show that there is no evidence at the 5% significance level that Samir's mean journey time has increased. **[5]**

**iii** State, with a reason, which one of the errors, Type I or Type II, might have been made in carrying out the test in part **ii**. **[2]**

*Cambridge International AS & A Level Mathematics 9709 Paper 73 Q6 June 2012*

9 The management of a factory thinks that the mean time required to complete a particular task is 22 minutes. The times, in minutes, taken by employees to complete this task have a normal distribution with mean $\mu$ and standard deviation 3.5. An employee claims that 22 minutes is not long enough for the task. In order to investigate this claim, the times for a random sample of 12 employees are used to test the null hypothesis $\mu = 22$ against the alternative hypothesis $\mu > 22$ at the 5% significance level.

i Show that the null hypothesis is rejected in favour of the alternative hypothesis if $\bar{x} > 23.7$ (correct to 3 significant figures), where $\bar{x}$ is the sample mean. **[3]**

ii Find the probability of a Type II error given that the actual mean time is 25.8 minutes. **[4]**

*Cambridge International AS & A Level Mathematics 9709 Paper 71 Q5 November 2011*

10 A doctor wishes to investigate the mean fat content in low-fat burgers. He takes a random sample of 15 burgers and sends them to a laboratory where the mass, in grams, of fat in each burger is determined. The results are as follows.

9 7 8 9 6 11 7 9 8 9 8 10 7 9 9

Assume that the mass, in grams, of fat in low-fat burgers is normally distributed with mean $\mu$ and that the population standard deviation is 1.3.

i Calculate a 99% confidence interval for $\mu$. **[4]**

ii Explain whether it was necessary to use the Central Limit Theorem in the calculation in part **i**. **[2]**

iii The manufacturer claims that the mean mass of fat in burgers of this type is 8 g. Use your answer to part **i** to comment on this claim. **[2]**

*Cambridge International AS & A Level Mathematics 9709 Paper 72 Q4 June 2011*

11 The masses of sweets produced by a machine are normally distributed with mean $\mu$ grams and standard deviation 1.0 grams. A random sample of 65 sweets produced by the machine has a mean mass of 29.6 grams.

i Find a 99% confidence interval for $\mu$. **[3]**

The manufacturer claims that the machine produces sweets with a mean mass of 30 grams.

ii Use the confidence interval found in part **i** to draw a conclusion about this claim. **[2]**

iii Another random sample of 65 sweets produced by the machine is taken. This sample gives a 99% confidence interval that leads to a different conclusion from that found in part **ii**. Assuming that the value of $\mu$ has not changed, explain how this can be possible. **[1]**

*Cambridge International AS & A Level Mathematics 9709 Paper 73 Q3 November 2010*

12 A random sample of $n$ people were questioned about their internet use. 87 of them had a high-speed internet connection. A confidence interval for the population proportion having a high-speed internet connection is $0.1129 < p < 0.1771$.

i Write down the mid-point of this confidence interval and hence find the value of $n$. **[3]**

ii This interval is an $\alpha$% confidence interval. Find $\alpha$. **[4]**

*Cambridge International AS & A Level Mathematics 9709 Paper 71 Q2 June 2010*

**13** The masses of packets of cornflakes are normally distributed with standard deviation 11 g. A random sample of 20 packets was weighed and found to have a mean mass of 746 g.

**i** Test at the 4% significance level whether there is enough evidence to conclude that the population mean mass is less than 750 g. **[4]**

**ii** Given that the population mean mass actually is 750 g, find the smallest possible sample size, $n$, for which it is at least 97% certain that the mean mass of the sample exceeds 745 g. **[4]**

*Cambridge International AS & A Level Mathematics 9709 Paper 72 Q5 November 2009*

## CROSS-TOPIC REVIEW EXERCISE 2

 **1** The time to failure, in years, for two types of kettle can be modelled by the continuous random variables $X$ and $Y$ which, respectively, have probability density functions as follows:

$$f(x) = \begin{cases} \dfrac{x}{8} & 0 \leqslant x \leqslant 4 \\ 0 & \text{otherwise} \end{cases} \qquad f(y) = \begin{cases} \dfrac{9}{4y^3} & 1 \leqslant y \leqslant 3 \\ 0 & \text{otherwise} \end{cases}$$

Show that the probability of failure by time $t$ is the same for both $X$ and $Y$ if $t$ satisfies the equation $t^4 - 18t^2 + 18 = 0$ and verify that this time is just over 1 year. **[7]**

**2** A continuous random variable $X$ has probability density function given by:

$$f(x) = \begin{cases} 0.25 & 4 \leqslant x \leqslant 8 \\ 0 & \text{otherwise} \end{cases}$$

**a** Sketch the graph of $y = f(x)$. **[2]**

**b** State the mean and use integration to find the variance. **[3]**

**c** The mean of a random sample of 40 observations of $X$ is denoted by $\bar{X}$. State the approximate distribution of $\bar{X}$, giving its parameters. **[3]**

**d** Find the value of $a$, where $P(\bar{X} < a) = 0.9$. **[3]**

 **3** Chakib cycles to college. He models his journey time, $T$ minutes, by the following probability density function:

$$f(t) = \begin{cases} \dfrac{1}{100}(25 - t) & 10 \leqslant t \leqslant 20 \\ 0 & \text{otherwise} \end{cases}$$

**a** Work out the mean and variance. **[7]**

Chakib finds that a random sample of 20 of his journey times has mean 12.4 minutes.

**b** Write down the approximate distribution of the sample mean for a sample of size 20. **[3]**

**c** Show that Chakib's model is not suitable. **[4]**

  **4 i** Give a reason for using a sample rather than the whole population in carrying out a statistical investigation. **[1]**

**ii** Tennis balls of a certain brand are known to have a mean height of bounce of 64.7 cm, when dropped from a height of 100 cm. A change is made in the manufacturing process and it is required to test whether this change has affected the mean height of bounce. 100 new tennis balls are tested and it is found that their mean height of bounce when dropped from a height of 100 cm is 65.7 cm and the unbiased estimate of the population variance is 15 cm$^2$.

**a** Calculate a 95% confidence interval for the population mean. **[3]**

**b** Use your answer to part **ii a** to explain what conclusion can be drawn about whether the change has affected the mean height of bounce. **[1]**

*Cambridge International AS & A Level Mathematics 9709 Paper 72 Q3 June 2016*

**5** The diameter, in cm, of pistons made in a certain factory is denoted by $X$, where $X$ is normally distributed with mean $\mu$ and variance $\sigma^2$. The diameters of a random sample of 100 pistons were measured, with the following results:

$$n = 100 \ \Sigma x = 208.7 \ \Sigma x^2 = 435.57$$

**i** Calculate unbiased estimates of $\mu$ and $\sigma^2$. **[3]**

The pistons are designed to fit into cylinders. The internal diameter, in cm, of the cylinders is denoted by $Y$, where $Y$ has an independent normal distribution with mean 2.12 and variance 0.000144. A piston will not fit into a cylinder if $Y - X < 0.01$.

**ii** Using your answers to part **i**, find the probability that a randomly chosen piston will not fit into a randomly chosen cylinder. **[6]**

*Cambridge International AS & A Level Mathematics 9709 Paper 73 Q7 November 2015*

**6** The marks, $x$, of a random sample of 50 students in a test were summarised as follows:

$$n = 50 \ \Sigma x = 1508 \ \Sigma x^2 = 51\,825$$

**i** Calculate unbiased estimates of the population mean and variance. **[3]**

**ii** Each student's mark is scaled using the formula $y = 1.5x + 10$. Find estimates of the population mean and variance of the scaled marks, $y$. **[3]**

*Cambridge International AS & A Level Mathematics 9709 Paper 73 Q4 June 2015*

**7** In a survey a random sample of 150 households in Nantville were asked to fill in a questionnaire about household budgeting.

**i** The results showed that 33 households owned more than one car. Find an approximate 99% confidence interval for the proportion of all households in Nantville with more than one car. **[4]**

**ii** The results also included the weekly expenditure on food, $x$ dollars, of the households. These were summarised as follows:

$$n = 150, \ \Sigma x = 19\,035, \ \text{and} \ \Sigma x^2 = 4\,054\,716$$

Find unbiased estimates of the mean and variance of the weekly expenditure on food of all households in Nantville. **[3]**

**iii** The government has a list of all the households in Nantville numbered from 1 to 9526. Describe briefly how to use random numbers to select a sample of 150 households from this list. **[3]**

*Cambridge International AS & A Level Mathematics 9709 Paper 72 Q4 November 2014*

**8** The number of hours that Mrs Hughes spends on her business in a week is normally distributed with mean $\mu$ and standard deviation 4.8. In the past the value of $\mu$ has been 49.5.

**i** Assuming that $\mu$ is still equal to 49.5, find the probability that in a random sample of 40 weeks the mean time spent on her business in a week is more than 50.3 hours. **[4]**

Following a change in her arrangements, Mrs Hughes wishes to test whether $\mu$ has decreased. She chooses a random sample of 40 weeks and notes that the total number of hours she spent on her business during these weeks is 1920.

ii a Explain why a one-tail test is appropriate. **[1]**

b Carry out the test at the 6% significance level. **[4]**

c Explain whether it was necessary to use the Central Limit theorem in part **ii b**. **[1]**

*Cambridge International AS & A Level Mathematics 9709 Paper 72 Q5 November 2014*

PS **9** Following a change in flight schedules, an airline pilot wished to test whether the mean distance that he flies in a week has changed. He noted the distances, $x$ km, that he flew in 50 randomly chosen weeks and summarised the results as follows.

$$n = 50,\ \Sigma x = 143\,300, \text{ and } \Sigma x^2 = 410\,900\,000$$

i Calculate unbiased estimates of the population mean and variance. **[3]**

ii In the past, the mean distance that he flew in a week was 2850 km. Test, at the 5% significance level, whether the mean distance has changed. **[5]**

*Cambridge International AS & A Level Mathematics 9709 Paper 71 Q3 November 2013*

**10** Each of a random sample of 15 students was asked how long they spent revising for an exam.

50 70 80 60 65 110 10 70 75 60 65 45 50 70 50

Assume that the times for all students are normally distributed with mean $\mu$ minutes and standard deviation 12 minutes.

i Calculate a 92% confidence interval for $\mu$. **[4]**

ii Explain what is meant by a 92% confidence interval for $\mu$. **[1]**

iii Explain what is meant by saying that a sample is 'random'. **[1]**

*Cambridge International AS & A Level Mathematics 9709 Paper 73 Q3 June 2013*

**11** In the past the weekly profit at a store had mean \$34 600 and standard deviation \$4500. Following a change of ownership, the mean weekly profit for 90 randomly chosen weeks was \$35 400.

i Stating a necessary assumption, test at the 5% significance level whether the mean weekly profit has increased. **[6]**

ii State, with a reason, whether it was necessary to use the Central Limit Theorem in part **i**. **[2]**

The mean weekly profit for another random sample of 90 weeks is found and the same test is carried out at the 5% significance level.

iii State the probability of a Type I error. **[1]**

iv Given that the population mean weekly profit is now \$36 500, calculate the probability of a Type II error. **[5]**

*Cambridge International AS & A Level Mathematics 9709 Paper 73 Q7 June 2013*

M **12** In order to obtain a random sample of people who live in her town, Jane chooses people at random from the telephone directory for her town.

i Give a reason why Jane's method will not give a random sample of people who live in the town. **[1]**

Jane now uses a valid method to choose a random sample of 200 people from her town and finds that 38 live in apartments.

**ii** Calculate an approximate 99% confidence interval for the proportion of all people in Jane's town who live in apartments. **[4]**

**iii** Jane uses the same sample to give a confidence interval of width 0.1 for this proportion. This interval is an $x$% confidence interval. Find the value of $x$. **[4]**

*Cambridge International AS & A Level Mathematics 9709 Paper 72 Q6 November 2012*

PS **13** The volumes of juice in bottles of Apricola are normally distributed. In a random sample of 8 bottles, the volumes of juice, in millilitres, were found to be as follows.

332 334 330 328 331 332 329 333

**i** Find unbiased estimates of the population mean and variance. **[3]**

A random sample of 50 bottles of Apricola gave unbiased estimates of 331 millilitres and 4.20 millilitres$^2$ for the population mean and variance respectively.

**ii** Use this sample of size 50 to calculate a 98% confidence interval for the population mean. **[3]**

**iii** The manufacturer claims that the mean volume of juice in all bottles is 333 millilitres. State, with a reason, whether your answer to part **ii** supports this claim. **[1]**

*Cambridge International AS & A Level Mathematics 9709 Paper 71 Q4 November 2011*

PS **14** Metal bolts are produced in large numbers and have lengths which are normally distributed with mean 2.62 cm and standard deviation 0.30 cm.



**i** Find the probability that a random sample of 45 bolts will have a mean length of more than 2.55 cm. **[3]**

**ii** The machine making these bolts is given an annual service. This may change the mean length of bolts produced but does not change the standard deviation. To test whether the mean has changed, a random sample of 30 bolts is taken and their lengths noted. The sample mean length is $m$ cm. Find the set of values of $m$ which result in rejection at the 10% significance level of the hypothesis that no change in the mean length has occurred. **[4]**

*Cambridge International AS & A Level Mathematics 9709 Paper 71 Q3 June 2010*

**15** There are 18 people in Millie's class. To choose a person at random she numbers the people in the class from 1 to 18 and presses the random number button on her calculator to obtain a 3-digit decimal. Millie then multiplies the first digit in this decimal by two and chooses the person corresponding to this new number. Decimals in which the first digit is zero are ignored.

**i** Give a reason why this is not a satisfactory method of choosing a person. **[1]**

Millie obtained a random sample of 5 people of her own age by a satisfactory sampling method and found that their heights in metres were 1.66, 1.68, 1.54, 1.65 and 1.57. Heights are known to be normally distributed with variance 0.0052 m$^2$.

**ii** Find a 98% confidence interval for the mean height of people of Millie's age. **[3]**

*Cambridge International AS & A Level Mathematics 9709 Paper 72 Q1 November 2009*

 **16** When Sunil travels from his home in England to visit his relatives in India, his journey is in four stages. The times, in hours, for the stages have independent normal distributions as follows:

Bus from home to the airport: N(3.75, 1.45)

Waiting in the airport: N(3.1, 0.785)

Flight from England to India: N(11, 1.3)

Car in India to relatives: N(3.2, 0.81)

i Find the probability that the flight time is shorter than the total time for the other three stages. **[6]**

ii Find the probability that, for 6 journeys to India, the mean time waiting in the airport is less than 4 hours. **[3]**

*Cambridge International AS & A Level Mathematics 9709 Paper 71 Q6 June 2009*

 **17** The times taken for the pupils in Ming's year group to do their English homework have a normal distribution with standard deviation 15.7 minutes. A teacher estimates that the mean time is 42 minutes. The times taken by a random sample of 3 students from the year group were 27, 35 and 43 minutes. Carry out a hypothesis test at the 10% significance level to determine whether the teacher's estimate for the mean should be accepted, stating the null and alternative hypotheses. **[5]**

*Cambridge International AS & A Level Mathematics 9709 Paper 7 Q2 November 2008*

 **18** Diameters of golf balls are known to be normally distributed with mean $\mu$ cm and standard deviation $\sigma$ cm. A random sample of 130 golf balls was taken and the diameters, $x$ cm, were measured. The results are summarised by $\sum x = 555.1$ and $\sum x^2 = 2371.30$.

i Calculate unbiased estimates of $\mu$ and $\sigma^2$. **[3]**

ii Calculate a 97% confidence interval for $\mu$. **[3]**

iii 300 random samples of 130 balls are taken and a 97% confidence interval is calculated for each sample. How many of these intervals would you expect **not** to contain $\mu$? **[1]**

*Cambridge International AS & A Level Mathematics 9709 Paper 7 Q4 November 2008*

 **19** The time in hours taken for clothes to dry can be modelled by the continuous random variable with probability density function given by:

$$f(t) = \begin{cases} k\sqrt{t} & 1 \leqslant t \leqslant 4, \\ 0 & \text{otherwise,} \end{cases}$$

where $k$ is a constant.

i Show that $k = \frac{3}{14}$. **[3]**

ii Find the mean time taken for clothes to dry. **[4]**

iii Find the median time taken for clothes to dry. **[3]**

iv Find the probability that the time taken for clothes to dry is between the mean time and the median time. **[2]**

*Cambridge International AS & A Level Mathematics 9709 Paper 7 Q7 November 2008*

**20** A magazine conducted a survey about the sleeping time of adults. A random sample of 12 adults was chosen from the adults travelling to work on a train.

**i** Give a reason why this is an unsatisfactory sample for the purposes of the survey. **[1]**

**ii** State a population for which this sample would be satisfactory. **[1]**

A satisfactory sample of 12 adults gave numbers of hours of sleep as shown below.

4.6 6.8 5.2 6.2 5.7 7.1 6.3 5.6 7.0 5.8 6.5 7.2

**iii** Calculate unbiased estimates of the mean and variance of the sleeping times of adults. **[3]**

*Cambridge International AS & A Level Mathematics 9709 Paper 7 Q1 June 2008*

## PRACTICE EXAM-STYLE PAPER

**Time allowed is 1 hour and 15 minutes (50 marks).**

1 Telephone calls to a company are answered within 3 minutes; if they are not answered within 3 minutes, they are disconnected. The length of time before calls are answered, $T$ minutes, is modelled using the probability density function:

$$f(t) = \begin{cases} kt(10 - 3t) & 0.5 \leqslant t \leqslant 3 \\ 0 & \text{otherwise} \end{cases}$$

where $k$ is a constant.

a Sketch the graph of $y = f(t)$ and show that $k = \frac{8}{135}$. **[5]**

b i Find the probability that a randomly chosen caller waits longer than 2 minutes for their call to be answered. **[2]**

ii For ten randomly chosen callers, find the probability that no more than one caller waits for longer than 2 minutes for their call to be answered. You may assume that the waiting times are independent. **[3]**

c Find the mean waiting time. **[3]**

d Explain whether or not the probability density function for $T$ is realistic. **[1]**

2 The number, $X$, of breakdowns per week of escalators in a large shopping mall has a Poisson distribution with mean 0.9.

a Find the probability that there will be:

i exactly three breakdowns during one week

ii exactly three breakdowns each week for three successive weeks

iii at most two breakdowns in a particular week. **[6]**

b Find the probability that during a particular four-week period there will be no breakdowns. **[2]**

c After the maintenance contract for the escalators is given to a new company, it is found there is one breakdown in a four-week period. Perform a 10% significance test to determine whether the average number of breakdowns has decreased. **[5]**

3 A fair coin is flipped 12 times. Heads appears two times. In a significance test for bias towards tails, the calculated probability results in a Type I error. Find the minimum level of significance of the test, giving your answer as an integer. **[4]**

4 The maker of a certain brand of chocolate sells boxes of individually wrapped chocolates. Weights of chocolates follow a normal distribution with mean 17 g and standard deviation 2 g. Weights of wrapping have an independent normal distribution with mean 4.5 g and standard deviation 0.3 g. Eight of these individually wrapped chocolates are placed together in a box. The weight of the box is 50 g. Find the probability that a randomly chosen box of chocolates weighs less than 230 g. **[5]**

5 Records from a physiotherapy clinic show that in one year the length of time for which patients are treated during a visit is normally distributed with mean 32.2 minutes and standard deviation 3.9.

a Calculate an interval within which 95% of the times will lie. **[3]**

b The following year a random sample of ten physiotherapy patients' visits have the following times (in minutes):

28.2 27.6 32.4 33.1 35.0 26.9 28.5 25.4 32.8 31.7

Assuming the times are normally distributed and the standard deviation remains unchanged from the previous year, use these times to calculate a 90% confidence interval for the population mean. **[4]**

6 A manufacturer claims the average lifetime of a particular battery it produces for smoke alarms is 76 weeks with a standard deviation of 4 weeks. To test this claim a random sample of 40 batteries is tested and found to have a mean lifetime of 74.6 weeks.

Stating clearly any assumptions you make, and using a 2% level of significance, test the claim made by the manufacturer. **[7]**

## THE STANDARD NORMAL DISTRIBUTION FUNCTION

If $Z$ is normally distributed with mean 0 and variance 1, the table gives the value of $\Phi(z)$ for each value of $z$, where

$$\Phi(z) = P(Z \leqslant z).$$

Use $\Phi(-z) = 1 - \Phi(z)$ for negative values of $z$.

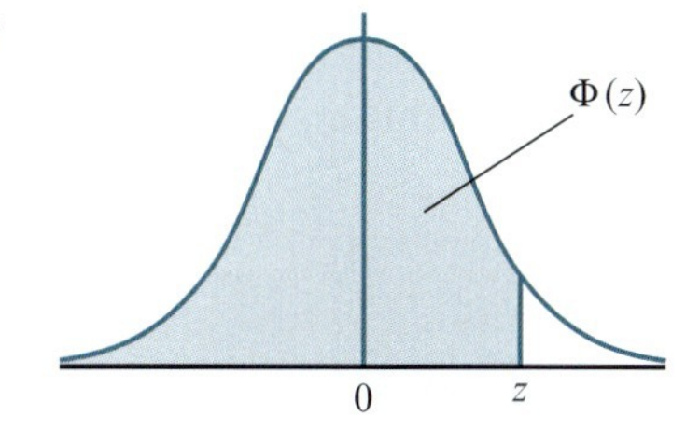


| z | 0 | 1 | 2 | 3 | 4 | 5 | 6 | 7 | 8 | 9 | 1 | 2 | 3 | 4 | 5 | 6 | 7 | 8 | 9 |
|---|---|---|---|---|---|---|---|---|---|---|---|---|---|---|---|---|---|---|---|
| | | | | | | | | | | | | | | | ADD | | | | |
| 0.0 | 0.5000 | 0.5040 | 0.5080 | 0.5120 | 0.5160 | 0.5199 | 0.5239 | 0.5279 | 0.5319 | 0.5359 | 4 | 8 | 12 | 16 | 20 | 24 | 28 | 32 | 36 |
| 0.1 | 0.5398 | 0.5438 | 0.5478 | 0.5517 | 0.5557 | 0.5596 | 0.5636 | 0.5675 | 0.5714 | 0.5753 | 4 | 8 | 12 | 16 | 20 | 24 | 28 | 32 | 36 |
| 0.2 | 0.5793 | 0.5832 | 0.5871 | 0.5910 | 0.5948 | 0.5987 | 0.6026 | 0.6064 | 0.6103 | 0.6141 | 4 | 8 | 12 | 15 | 19 | 23 | 27 | 31 | 35 |
| 0.3 | 0.6179 | 0.6217 | 0.6255 | 0.6293 | 0.6331 | 0.6368 | 0.6406 | 0.6443 | 0.6480 | 0.6517 | 4 | 7 | 11 | 15 | 19 | 22 | 26 | 30 | 34 |
| 0.4 | 0.6554 | 0.6591 | 0.6628 | 0.6664 | 0.6700 | 0.6736 | 0.6772 | 0.6808 | 0.6844 | 0.6879 | 4 | 7 | 11 | 14 | 18 | 22 | 25 | 29 | 32 |
| 0.5 | 0.6915 | 0.6950 | 0.6985 | 0.7019 | 0.7054 | 0.7088 | 0.7123 | 0.7157 | 0.7190 | 0.7224 | 3 | 7 | 10 | 14 | 17 | 20 | 24 | 27 | 31 |
| 0.6 | 0.7257 | 0.7291 | 0.7324 | 0.7357 | 0.7389 | 0.7422 | 0.7454 | 0.7486 | 0.7517 | 0.7549 | 3 | 7 | 10 | 13 | 16 | 19 | 23 | 26 | 29 |
| 0.7 | 0.7580 | 0.7611 | 0.7642 | 0.7673 | 0.7704 | 0.7734 | 0.7764 | 0.7794 | 0.7823 | 0.7852 | 3 | 6 | 9 | 12 | 15 | 18 | 21 | 24 | 27 |
| 0.8 | 0.7881 | 0.7910 | 0.7939 | 0.7967 | 0.7995 | 0.8023 | 0.8051 | 0.8078 | 0.8106 | 0.8133 | 3 | 5 | 8 | 11 | 14 | 16 | 19 | 22 | 25 |
| 0.9 | 0.8159 | 0.8186 | 0.8212 | 0.8238 | 0.8264 | 0.8289 | 0.8315 | 0.8340 | 0.8365 | 0.8389 | 3 | 5 | 8 | 10 | 13 | 15 | 18 | 20 | 23 |
| 1.0 | 0.8413 | 0.8438 | 0.8461 | 0.8485 | 0.8508 | 0.8531 | 0.8554 | 0.8577 | 0.8599 | 0.8621 | 2 | 5 | 7 | 9 | 12 | 14 | 16 | 19 | 21 |
| 1.1 | 0.8643 | 0.8665 | 0.8686 | 0.8708 | 0.8729 | 0.8749 | 0.8770 | 0.8790 | 0.8810 | 0.8830 | 2 | 4 | 6 | 8 | 10 | 12 | 14 | 16 | 18 |
| 1.2 | 0.8849 | 0.8869 | 0.8888 | 0.8907 | 0.8925 | 0.8944 | 0.8962 | 0.8980 | 0.8997 | 0.9015 | 2 | 4 | 6 | 7 | 9 | 11 | 13 | 15 | 17 |
| 1.3 | 0.9032 | 0.9049 | 0.9066 | 0.9082 | 0.9099 | 0.9115 | 0.9131 | 0.9147 | 0.9162 | 0.9177 | 2 | 3 | 5 | 6 | 8 | 10 | 11 | 13 | 14 |
| 1.4 | 0.9192 | 0.9207 | 0.9222 | 0.9236 | 0.9251 | 0.9265 | 0.9279 | 0.9292 | 0.9306 | 0.9319 | 1 | 3 | 4 | 6 | 7 | 8 | 10 | 11 | 13 |
| 1.5 | 0.9332 | 0.9345 | 0.9357 | 0.9370 | 0.9382 | 0.9394 | 0.9406 | 0.9418 | 0.9429 | 0.9441 | 1 | 2 | 4 | 5 | 6 | 7 | 8 | 10 | 11 |
| 1.6 | 0.9452 | 0.9463 | 0.9474 | 0.9484 | 0.9495 | 0.9505 | 0.9515 | 0.9525 | 0.9535 | 0.9545 | 1 | 2 | 3 | 4 | 5 | 6 | 7 | 8 | 9 |
| 1.7 | 0.9554 | 0.9564 | 0.9573 | 0.9582 | 0.9591 | 0.9599 | 0.9608 | 0.9616 | 0.9625 | 0.9633 | 1 | 2 | 3 | 4 | 4 | 5 | 6 | 7 | 8 |
| 1.8 | 0.9641 | 0.9649 | 0.9656 | 0.9664 | 0.9671 | 0.9678 | 0.9686 | 0.9693 | 0.9699 | 0.9706 | 1 | 1 | 2 | 3 | 4 | 4 | 5 | 6 | 6 |
| 1.9 | 0.9713 | 0.9719 | 0.9726 | 0.9732 | 0.9738 | 0.9744 | 0.9750 | 0.9756 | 0.9761 | 0.9767 | 1 | 1 | 2 | 2 | 3 | 4 | 4 | 5 | 5 |
| 2.0 | 0.9772 | 0.9778 | 0.9783 | 0.9788 | 0.9793 | 0.9798 | 0.9803 | 0.9808 | 0.9812 | 0.9817 | 0 | 1 | 1 | 2 | 2 | 3 | 3 | 4 | 4 |
| 2.1 | 0.9821 | 0.9826 | 0.9830 | 0.9834 | 0.9838 | 0.9842 | 0.9846 | 0.9850 | 0.9854 | 0.9857 | 0 | 1 | 1 | 2 | 2 | 2 | 3 | 3 | 4 |
| 2.2 | 0.9861 | 0.9864 | 0.9868 | 0.9871 | 0.9875 | 0.9878 | 0.9881 | 0.9884 | 0.9887 | 0.9890 | 0 | 1 | 1 | 1 | 2 | 2 | 2 | 3 | 3 |
| 2.3 | 0.9893 | 0.9896 | 0.9898 | 0.9901 | 0.9904 | 0.9906 | 0.9909 | 0.9911 | 0.9913 | 0.9916 | 0 | 1 | 1 | 1 | 1 | 2 | 2 | 2 | 2 |
| 2.4 | 0.9918 | 0.9920 | 0.9922 | 0.9925 | 0.9927 | 0.9929 | 0.9931 | 0.9932 | 0.9934 | 0.9936 | 0 | 0 | 1 | 1 | 1 | 1 | 1 | 2 | 2 |
| 2.5 | 0.9938 | 0.9940 | 0.9941 | 0.9943 | 0.9945 | 0.9946 | 0.9948 | 0.9949 | 0.9951 | 0.9952 | 0 | 0 | 0 | 1 | 1 | 1 | 1 | 1 | 1 |
| 2.6 | 0.9953 | 0.9955 | 0.9956 | 0.9957 | 0.9959 | 0.9960 | 0.9961 | 0.9962 | 0.9963 | 0.9964 | 0 | 0 | 0 | 0 | 1 | 1 | 1 | 1 | 1 |
| 2.7 | 0.9965 | 0.9966 | 0.9967 | 0.9968 | 0.9969 | 0.9970 | 0.9971 | 0.9972 | 0.9973 | 0.9974 | 0 | 0 | 0 | 0 | 0 | 1 | 1 | 1 | 1 |
| 2.8 | 0.9974 | 0.9975 | 0.9976 | 0.9977 | 0.9977 | 0.9978 | 0.9979 | 0.9979 | 0.9980 | 0.9981 | 0 | 0 | 0 | 0 | 0 | 0 | 0 | 1 | 1 |
| 2.9 | 0.9981 | 0.9982 | 0.9982 | 0.9983 | 0.9984 | 0.9984 | 0.9985 | 0.9985 | 0.9986 | 0.9986 | 0 | 0 | 0 | 0 | 0 | 0 | 0 | 0 | 0 |

**Critical values for the normal distribution**

The table gives the value of $z$ such that $P(Z \leqslant z) = p$, where $Z \sim N(0, 1)$.

| $p$ | 0.75 | 0.90 | 0.95 | 0.975 | 0.99 | 0.995 | 0.9975 | 0.999 | 0.9995 |
|---|---|---|---|---|---|---|---|---|---|
| $z$ | 0.674 | 1.282 | 1.645 | 1.960 | 2.326 | 2.576 | 2.807 | 3.090 | 3.291 |

# Answers

## 1 Hypothesis testing

### Prerequisite knowledge

1 a 0.149 b 0.813 c 0.00502

2 a 0.0393 to 3 s.f. b 0.0802 to 3 s.f. c 0.381

3 a 0.692 b 0.960 to 3 s.f. c 0.440

4 a 0.122 b 0.798

5 $X \sim N(32, 19.2)$

a 0.848 b 0.634

6 $X \sim N(66, 29.7)$

a 0.205 b 0.323

### Exercise 1A

1 a Let $X$ be the number of nature programmes, $X \sim B(20, 0.25)$; $H_0$: $p = 0.25$; $H_1$: $p < 0.25$

b $P(X \leqslant 2) = 0.0913$. As this is less than 10%, reject $H_0$. There is insufficient evidence to accept the television channel's claim.

c 0.0913 is greater than 5%, therefore accept $H_0$. There is insufficient evidence to reject the claim that 25% of programmes are nature programmes.

2 Let $X$ be the number of footballers who cannot explain the offside rule, $X \sim B(15, 0.4)$; $H_0$: $p = 0.4$; $H_1$: $p < 0.4$

$P(X \leqslant 2) = 0.0271$. As this is less than 5%, reject $H_0$. There is some evidence to support the claim that the number of professional footballers who cannot explain the offside rule is less than 40%.

3 a Let $X$ be the number of sheep deficient in a given mineral, $X \sim B(80, 0.3)$; $H_0$: $p = 0.3$; $H_1$: $p < 0.3$

$P(X \leqslant 19) = 0.136$. As this is greater than 10%, accept $H_0$. There is insufficient evidence to support the claim that the percentage of sheep deficient in a particular mineral has decreased.

b 18

4 a $P(X \geqslant 7) = 0.0898$. As this is greater than 5%, accept $H_0$. There is insufficient evidence to support the claim that the coin is biased towards tails.

b $Y \sim B(9, 0.5)$; $H_0$: $p = 0.5$; $H_1$: $p < 0.5$

$P(Y \leqslant 2) = 0.0898 > 5\%$; accept $H_0$.

There is insufficient evidence to support the claim that the coin is biased towards tails.

c $X \sim N(90, 45)$; $H_0$: $\mu = 90$; $H_1$: $\mu > 90$

$P(X \geqslant 102) = 0.0432$. As this is less than 5%, reject $H_0$. There is some evidence to support the claim that the coin is biased towards tails.

5 $X \sim B(120, 0.25)$; $H_0$: $p = 0.25$; $H_1$: $p > 0.25$ or $X \sim N(30, 22.5)$; $H_0$: $\mu = 30$; $H_1$: $\mu > 30$

$P(X \geqslant 40) = 0.0226$. As this is greater than 2%, accept $H_0$. There is insufficient evidence to support the claim that the spinner is biased towards blue.

6 $X \sim B(17, 0.6)$; $H_0$: $p = 0.6$; $H_1$: $p > 0.6$

Test statistic is $P(X \geqslant 15) = 0.0123$.

Significance level is 2%.

### Exercise 1B

1 a $X \sim B(20, 0.25)$; $H_0$: $p = 0.25$; $H_1$: $p \neq 0.25$

b $P(X \leqslant 2) = 0.0913$. As this is greater than 5%, accept $H_0$. There is insufficient evidence to reject the claim that 25% of programmes are nature programmes.

2 $X \sim B(15, 0.4)$; $H_0$: $p = 0.4$; $H_1$: $p \neq 0.4$

$P(X \leqslant 2) = 0.0271$. As this is greater than 2.5%, accept $H_0$. There is insufficient evidence to reject the claim that 40% of professional footballers cannot explain the offside rule.

3 a $X \sim B(16, 0.8)$; $H_0$: $p = 0.8$; $H_1$: $p \neq 0.8$

$P(X \geqslant 15) = 0.141$. As this is greater than 5%, accept $H_0$. There is insufficient evidence to reject the claim that 80% of Americans believe in horoscopes.

Critical value = 16

b $X \sim B(160, 0.8)$; $H_0$: $p = 0.8$; $H_1$: $p \neq 0.8$ or $X \sim N(128, 25.6)$; $H_0$: $\mu = 128$; $H_1$: $\mu \neq 128$

$P(X \leqslant 116) = 0.0115$. As this is less than 2.5%, reject $H_0$. There is sufficient evidence to reject the claim that 80% of Americans believe in horoscopes.

4 a $X \sim B(180, 0.12)$; $H_0$: $p = 0.12$; $H_1$: $p \neq 0.12$ or $X \sim N(21.6, 19.008)$; $H_0$: $\mu = 21.6$; $H_1$: $\mu \neq 21.6$

$P(X \leqslant 13) = 0.0316$

In a two tailed test we require $2 \times$ probability < significance level to reject $H_0$.

$2 \times 0.0316 = 0.0632 = 6.32\%$.

The minimum integer significance level is 7%; Magnus concludes there is insufficient evidence to support the claim.

Since $P(X \leqslant 14) = 0.0517$ and $2 \times 0.0517 = 0.1034 = 10.34\%$, if the critical value is 13 then an integer significance level can be 7, 8, 9 or 10%.

b $0.0316 < 7\%$ or $8\%$ or $9\%$ or $10\%$, therefore reject $H_0$. There is some evidence to suggest that a lower percentage of people can spell onomatopoeia.

5 a $X \sim B(20, 0.3)$; $H_0$: $p = 0.3$; $H_1$: $p \neq 0.3$

$P(X \leqslant 3) = 0.107$. As this is greater than $5\%$, accept $H_0$. There is insufficient evidence to reject the claim that 30% are red.

Critical value = 2

b $H_0$: $p = 0.3$; $H_1$: $p < 0.3$

No, Ginny's conclusion will be the same.

6 $X \sim B(600, 0.01)$; $H_0$: $p = 0.01$; $H_1$: $p \neq 0.01$ or $X \sim N(6, 5.94)$; $H_0$: $\mu = 6$; $H_1$: $\mu \neq 6$

$P(X \leqslant 11) = 0.988$. $1 - 0.988 < 2\%$, reject $H_0$. There is insufficient evidence to accept the claim.

## Exercise 1C

1 a $X \sim B(12, 0.4)$

b $H_0$: $p = 0.4$; $H_1$: $p < 0.4$

c $P(X \leqslant a) < 5\%$

$P(X \leqslant 1) = 0.0196 < 5\%$, whereas $P(X \leqslant 2) = 0.0835 > 5\%$.
So the rejection region is $X \leqslant 1$



d So you reject the archer's claim if $X = 1$.

2 a $X \sim B(16, 0.8)$

b $H_0$: $p = 0.8$; $H_1$: $p > 0.8$

c $P(X \geqslant a) < 10\%$

$P(X \geqslant 15) = 0.1407 > 10\%$, whereas $P(X \geqslant 16) = 0.0281 < 10\%$. So the rejection region is $X > 16$.

d So the probability of a Type I error is 0.0281.

3 $X \sim B(300, 0.2)$; $H_0$: $p = 0.2$; $H_1$: $p < 0.2$

$X \sim N(60, 48)$; $H_0$: $\mu = 60$; $H_1$: $\mu < 60$

$P(X \leqslant a - 0.5) < 5\%$

$$\frac{a - 0.5 - 60}{\sqrt{48}} = -1.645$$

$$a = 49.1$$

49 teenagers

4 a $H_0$: $\mu = 1.8$; $H_1$: $\mu > 1.8$

b $$1 - \Phi\left(\frac{2.25 - 1.8}{0.32}\right) = 1 - \Phi(1.406) = 1 - 0.9200 = 0.0800$$

c $$\Phi\left(\frac{2.25 - 2.4}{0.32}\right) = \Phi(-0.46875) = 1 - 0.6804 = 0.320$$

5 a Die is considered to be biased when it is in fact fair.

$X \sim B(20, 0.125)$

$P(X \geqslant 4) = 1 - P(X \leqslant 3) = 1 - 0.765 = 0.235$

b Die is considered to be fair when it is in fact biased.

Need to know the probability of rolling a 1 with a biased die.

## End-of-chapter review exercise 1

1 a $X \sim B(25, 0.16)$; $H_0$: $p = 0.16$; $H_1$: $p < 0.16$

$P(X \leqslant 1) = 0.0737 > 5\%$, so accept $H_0$.

b $X \sim B(150, 0.16)$

$X \sim N(24, 20.16)$; $H_0$: $\mu = 24$; $H_1$: $\mu < 24$

$$\Phi\left(\frac{18.5 - 24}{\sqrt{20.16}}\right) = \Phi(-1.225) = 1 - 0.8897 = 0.1103 > 5\%$$

So accept $H_0$.

There is no evidence to suggest that the new clay is more reliable.

2 $X \sim B(1000, 0.2)$; $X \sim N(200, 160)$; $H_0$: $\mu = 200$; $H_1$: $\mu > 200$

$$\Phi\left(\frac{217.5 - 200}{\sqrt{160}}\right) = \Phi(1.383) = 0.9167 < 95\%$$

So accept $H_0$.

There is no evidence to suggest the fens will flood more often.

3 a $X \sim N(800, 50^2)$; $H_0$: $\mu = 800$; $H_1$: $\mu \neq 800$

$$\Phi\left(\frac{720 - 800}{50}\right) = 0.0548$$

The light bulb lasting 720 hours is fewer hours than claimed, but the null hypothesis is accepted; if the light bulb lasted for this time, statistically you could not say the manufacturer's claim is unjustified.

$$1 - \Phi\left(\frac{920 - 800}{50}\right) = 0.00820$$

The light bulb lasting 920 hours statistically lasts longer than the manufacturer's claims; you would reject $H_0$ and accept $H_1$. In this case, however, your light bulb is longer-lasting than expected, so you may not want to challenge the manufacturer's claim.

b 702 hours

4 $X \sim B(40, 0.08)$

$P(X \leqslant 0) = 0.0356 < 10\%$

$P(X \leqslant 1) = 0.1594 > 10\%$

Critical region $X \leqslant 0$

Critical value $X = 0$

5 a $X \sim B(850, 0.012)$; $X \sim N(10.2, 10.0776)$;
$H_0: \mu = 10.2$; $H_1: \mu > 10.2$
$1 - \Phi\left(\frac{14.5 - 10.2}{\sqrt{10.0776}}\right) = 1 - \Phi(1.355) = 1 - 0.9123 > 2\%$
So accept $H_0$.
There is no evidence to suggest there is a greater proportion of red-haired people in Scotland.

b $1 - \Phi\left(\frac{x - 0.5 - 10.2}{\sqrt{10.0776}}\right) < 0.02$
$\frac{x - 0.5 - 10.2}{\sqrt{10.0776}} > 2.054$
Critical value $= 18$

6 i $H_0: p = 0.3$; $H_1: p < 0.3$
ii Probability of Type I error $= 0.0355$
iii There is insufficient evidence to suggest the proportion of packets containing gifts is less than 30%.

7 i $H_0: p = \frac{1}{8}$; $H_1: p > \frac{1}{8}$
ii 0.120 iii 12%

8 i There is insufficient evidence to support the claim.
ii Normal; $\mu = 200$; $\sigma^2 = 160$ or $\sigma = \sqrt{160}$
iii Type II error = accepting the hypothesis that the machine produces the correct number of 5s when it does not.

9 i Fewer than 2 samples contain a free gift.
ii 0.0243
iii Two is not in the critical region, so there is insufficient evidence to reject the claim.

10 There is insufficient evidence to support the claim.

11 i $H_0: p = \frac{1}{6}$; $H_1: p > \frac{1}{6}$
ii 0.0697
iii A Type II error occurs if the die is biased towards six, but less than 4 sixes are thrown, so there is no evidence of bias.
iv 0.172 or $\frac{11}{64}$

12 There is insufficient evidence to support the claim that women in Jakarta have smaller fingers.

# 2 The Poisson distribution

## Prerequisite knowledge

1 a 0.000786 b 0.376
2 a 0.999 b 0.215
3 a 0.135 b 0.0273 c 0.000123
4 $H_0: \mu = 42$; $H_1: \mu \neq 42$; $\Phi\left(\frac{45 - 42}{\sqrt{8}}\right) = \Phi(1.061) =$ 0.1444

## Exercise 2A

1 a 0.156 b 0.238 c 0.982
2 a 0.149 b 0.970
3 a 0.0471 b 0.809 c 0.191
4 a 0.125
b Occurrence of potholes may be affected by factors such as materials used, quality of work, weather, etc.
5 a i Calls occur singly, independently and at random.
ii Unlikely as more than one person may call to report a large fire; hence, calls may not always be independent.
b 0.0153
6 a i 0.134 ii 0.928
b Choose some bars of the same hazelnut chocolate at random; count the number of whole hazelnuts; draw a frequency table and calculate mean and variance.
7 a Mean 1.625 or 1.63; variance 1.61; mean and variance are similar.
b 13, 20, 17, 9, 4, 1, 0, 0

## Exercise 2B

1 a 0.113 b 0.000499
2 0.0907
3 Flaws occur independently and at random.
a 0.000335 b 0.191
4 a 0.958
b Possible accident or road closure, or the random 20 minutes happened to be in the middle of the night.
c Traffic flow different, cars not independent.
5 a 0.0183 b 0.268
6 a 0.406 b 0.0916
7 a Orders are placed independently and at random.
b i 0.287 ii 0.358
8 a Emails are received independently and at random.
b A faulty product may generate lots of emails.
c People unlikely to complain at night.
d i 0.129 ii 0.0000533

## Exercise 2C

1 a 0.647 b 0.353
2 a 0.964 b 0.430

**3** 0.00449

**4** 0.937

**5** $X \sim \text{Po}(1.2)$; 0.966

**6** **a** 0.0656

**b** Accept 13 800 or 13 816.

**7** **a** $n$ large, $p$ small

**b** 0.294 **c** 0.259

## Exercise 2D

**1** **a** 0.876 **b** 0.409 **c** 0.0552

**2** **a** 0.144 **b** 0.167

**3** 0.186

**4** **a** 0.119

**b** No; the reason for delay (e.g. weather/accident) can affect delays on subsequent days.

**5** 0.964

**6** 50

**7** 51

**8** **a** 0.0668

**b** $\Phi\left(\frac{38.5-25}{\sqrt{25}}\right)-\left(1-\Phi\left(\frac{11.5-25}{\sqrt{25}}\right)\right)$

$= 0.9965-(1-0.9965) = 0.993$



## Exercise 2E

**1** $H_0: \lambda = 62$ or $\mu = 62$

$H_1: \lambda > 62$ or $\mu > 62$

Probability $= 0.170$

Accept $H_0$.

**2** $H_0: \lambda = 7$; $H_1: \lambda < 7$

Probability $= 0.173$

Accept $H_0$.

**3** $H_0: \lambda = 39$ or $\mu = 39$; $H_1: \lambda < 39$ or $\mu < 39$

Probability $= 0.0464$

Reject $H_0$.

**4** **a** $H_0: \lambda = 64$ or $\mu = 64$; $H_1: \lambda \neq 64$ or $\mu \neq 64$

Probability $= 0.0349$

Accept $H_0$.

**b** 47 or fewer, and 80 or more.

## End-of-chapter review exercise 2

**1** **a** 0.195 **b** 0.762

**c** 0.122 **d** 0.809

**2** **a** Mean = 3.425; variance = 3.21. These are fairly close, using a Poisson distribution is appropriate.

Could also use the mean to calculate expected frequencies for comparison.

**b** 0.433

**3** **a** 0.0796 **b** 0.0687 **c** 0.385

**4** $H_0: \lambda = 3$; $H_1: \lambda > 3$

Probability $(X \geqslant 5) = 0.185 > 5\%$

This is not in the critical region, so accept $H_0$.

**5** 0.353

**6** **i** 0.135

**ii** **a** 0.744 **b** 0.870

**iii** 6

**7** **i** Constant mean rate **ii** 0.0922

**iii** 0.821 **iv** 0.0723

**8** **i** 0.469 **ii** 2.15 minutes

**9** **i** Customers arrive independently.

**ii** 0.161 **iii** 0.570 **iv** 0.869

**10** **i** Constant average rate of goals scored; goals scored independently.

**ii** 0.259 **iii** 0.164

**11** **i** $P(X \geq 5) = 0.0959 > 0.05$ There is insufficient evidence to suggest mean sales have increased.

**ii** 0.0357

**iii** Mean sales = 0.8 per week but 6 or more sales in 3 weeks leads to rejecting mean = 0.8.

**iv** The value of the new mean.

**12** **i** Type I error shows the number of white blood cells has decreased when in fact it has not decreased. Probability = 0.0342

**ii** There is insufficient evidence to show a decrease in the number of white blood cells.

**iii** 0.915

**13** **i** 0.143 **ii** 0.118 **iii** 0.0316

# 3 Linear combinations of random variables

## Prerequisite knowledge

**1** $E(X) = 2\frac{1}{6}$; $\text{Var}(X) = \frac{29}{36}$

**2** **a** 0.673 **b** 0.673

**3** **a** 0.119 **b** 0.0289

## Exercise 3A

**1** **a** 74 **b** 150

**2** **a** 15 **b** 19.2

3 a i 9.2 ii 7.2

b $a = 5, b = 20$ or $a = -5, b = 44$

4 a i 19 ii 20.8

b $a = 10, b = 30$ or $a = -10, b = -70$

5 a 8.5 b 15.25

6 a 11 b $46\frac{2}{3}$

7 a 25.8 b 310

8 $a = \frac{1}{\sigma}, b = \frac{\mu}{\sigma}$ or $a = \frac{-1}{\sigma}, b = \frac{-\mu}{\sigma}$

9 Mean = 76.1; standard deviation = 2.55

## Exercise 3B

1 a 7.2 b 19.68

2 a 18 b 18.3

3 a 39; 6.6 b −8; 3.5

4 a 22.5 b 14.6

5 a 20 b 4.3 c 7 d 16.2

6 a 3; 1.8 b 6; 7.2 c 6 d 3.6

7 Mean = 11; variance = $51\frac{1}{6}$; sd = 7.15

## Exercise 3C

1 a $A + 2B \sim N(66, 104)$

b $2B - A \sim N(46, 104)$

c $A + B + C \sim N(53, 38)$

2 a $2X - Y \sim N(23, 5^2)$ b 0.841

3 a 0.421 b 0.788

4 0.401

5 7020; 76.8

6 0.0455

7 0.607

8 a N(1.6, 0.0008) b N(1.6, 0.0128)

## Exercise 3D

1 a 2.1; 23.1 b 10.5; 10.5

c 23.1; 60.9 d **b** has a Poisson distribution.

2 a 12 b 0.000442

3 0.00470

4 a 0.128 b 0.0111

5 $T \sim Po(3)$; 0.448

6 a 0.0361 b 0.594

7 a $T \sim Po(3.7)$

b i 0.193 ii 0.285

8 0.175

9 Assume items are received at random.

a 0.215 b 0.228

10 0.704. It is necessary to assume that goals scored by each team are independent. Not reasonable since the teams are opponents.

## End-of-chapter review exercise 3

1 Mean = 63.7; standard deviation = 7.2

2 0.747

3 0.149

4 0.542

5 0.721

6 0.0356

7 i 0.889

ii 0.176

8 i 9.3 ii 27.9

iii Mean = −2.0; variance = 37.2

9 i Mean = 926 cents; sd = 105 cents

ii Mean = 5556 cents; sd = 257 cents

10 0.9405

11 i 0.823

ii Mean = 376 miles; sd = 39.0 miles

12 i $a = -6.32; b = 2$

ii a 1.87 b 0.699

# Cross-topic review exercise 1

1 a 0.951

b 0.00709

c $0.967 < 0.98$

There is insufficient evidence to reject the claim.

2 0.336

3 $\lambda + 2.326\sqrt{\lambda} - 55.5 = 0$

$\lambda = 40.7$

4 i 0.216 ii 0.973 iii 0.221

5 i 0.186 ii 0.257 iii 0.191

iv The officer will reject $H_0$.

6 i 0.0254 ii 0.983

7 i 0.256

ii 0.117

iii Type I: Luigi's belief is rejected, though it may be true.

**8** **i** 0.259 **ii** $e^{-2} \times \frac{2^r}{r!} = \frac{2}{3}e^{-2}$

$3 \times 2^3 = 4!$

**9** **i** 0.294 **ii** 5 **iii** 0.0753

**10** **i** 0.191 **ii** 0.112 **iii** 3

**11** **i** 0.202 **ii** 0.159

**12** **i** 0.0202 **ii** 0.972 **iii** 0.0311

**13** **i** 0.938 **ii** 0.993

**14** **i** 5 years = 60 months; 1 in 15 ≡ 4 in 60

**ii** 0 or 1

**iii** 0.0916

**iv** 1 lies in the rejection region; there is evidence that the new guitar strings last longer.

# 4 Continuous random variables

## Prerequisite knowledge

**1** 22

**2** 0.8

**3** 10

**4** 0.785

**5** $m = 0.5$



## Exercise 4A

**1** **a** $\int_0^2 kx^4 + \frac{1}{5}\,dx = 1$

$\left[\frac{kx^5}{5} + \frac{x}{5}\right]_0^2 = 1$

$\left(\frac{32k}{5} + \frac{2}{5}\right) = 1$

$k = \frac{3}{32}$

**b**

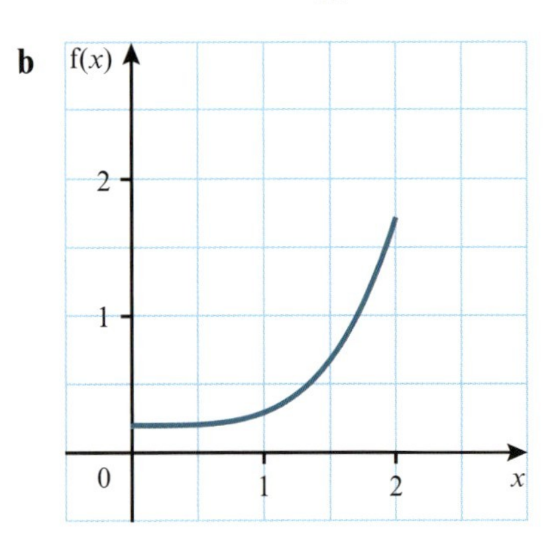

**c** $\frac{7}{32}$

**2** **a** $\int_2^4 \frac{x^3}{60}\,dx = \left[\frac{x^4}{240}\right]_2^4 = \frac{256}{240} - \frac{16}{240} = 1$

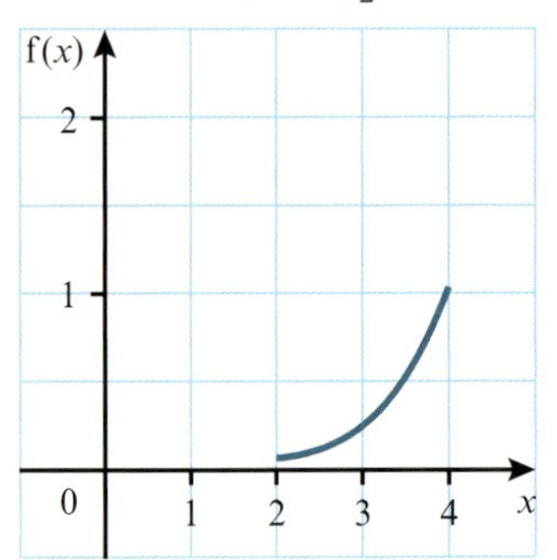

**b** $\frac{35}{48}$ **c** 0

**3** **a** $\int_0^2 \frac{3x^2}{32}\,dx + \int_2^5 \frac{1}{24}(13 - 2x)\,dx$

$= \left[\frac{x^3}{32}\right]_0^2 + \left[\frac{1}{24}(13x - x^2)\right]_2^5 = 1$

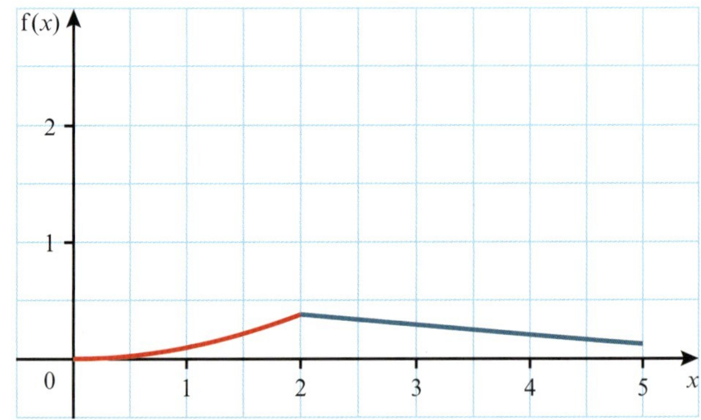

**b** **i** $\frac{1}{6}$

**ii** $\frac{53}{96}$ or 0.552

**4** **a** $\int_1^2 \frac{2(x-1)}{3}\,dx + \int_2^4 \frac{4-x}{3}\,dx$

$= \left[\frac{x^2 - 2x}{3}\right]_1^2 + \left[\frac{4x}{3} - \frac{x^2}{6}\right]_2^4 = 1$

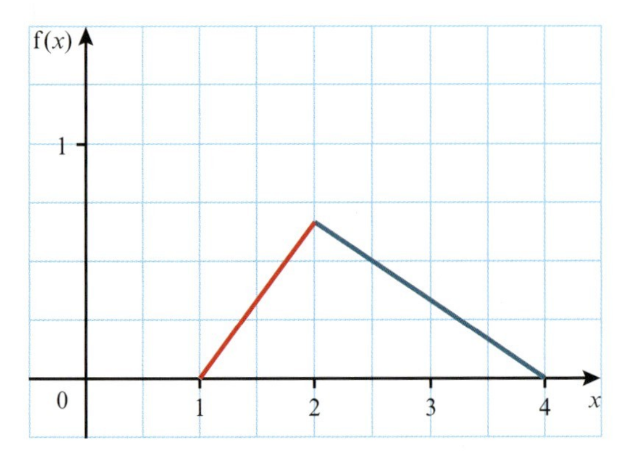

**b** $\frac{5}{6}$

**c** $k = 2$

5 1.25

6 $\frac{11}{40}$

7 0.0183

## Exercise 4B

1 a $\int_2^5 k(x-2)\,dx + \int_5^7 k(7-x)\,dx = 1$

b $\frac{4}{13}$ c $\frac{11}{13}$

d $2+\frac{\sqrt{26}}{2}$ e $2+\frac{\sqrt{65}}{5}$

2 a $\frac{16}{25}$

b Solve $m^2 - 10m + 12.5 = 0$; other root outside range.

c $\frac{5}{2}(\sqrt{3}-1)$

3 a Sketch or $\int_{10}^{30} \frac{1}{k}(40-t)\,dt = 1$

b $\frac{11}{32}$ or 0.34375

c $40 - 10\sqrt{5}$

d Likely to talk for shorter or longer periods.

4 a

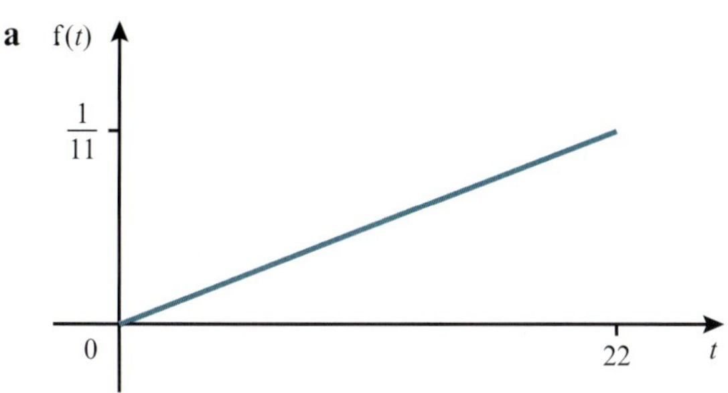


b 0.948 or $\frac{459}{484}$

c Either integrate with limits from 0 to 20 (gives probability 0.826 or $\frac{100}{141} > 80\%$) or attempt to find upper value with integral = 0.8, upper value 19.7 (3 significant figures).

5 a Graphically or $\int_0^{2.5} kx\,dx + \int_{2.5}^5 k(5-x)\,dx = 1$

b 0.6 or $\frac{3}{5}$

c

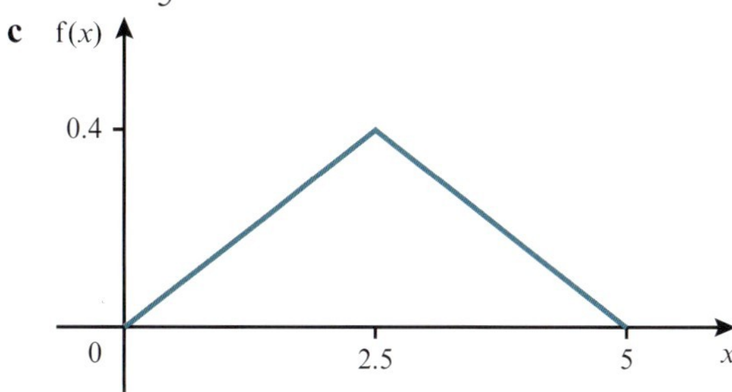


i Peak at 2.5

ii Mid point of symmetrical graph, 2.5

6 a 0.812 b 800

## Exercise 4C

1 Mean = 2.5; variance = 1.25

2 Mean = 2.81; variance = 1.13

3 a $k\left(\left[\frac{x^3}{3}\right]_0^2 + \left[6x - \frac{x^2}{2}\right]_2^6\right) = 1$

$k\left(\frac{8}{3} + 18 - 10\right) = 1$

$k = \frac{3}{32}$ as required

b Mean = 2.88; variance = 1.33

4 a $k\left[x^3 - 2x^2 + 2x\right]_3^5 = 1$

$k(85 - 15) = 1$

$k = \frac{1}{70}$ as required

b 4.19 kg c \$264

5 a 0.4 b 4.08 or 4

6 a Mean = $\frac{287}{150}$ or 1.91; sd = 0.522

b 0.510

c 600

7 a 0.741

b Mean = 5; variance = 25

## End-of-chapter review exercise 4

**1** **a**

**b** $\frac{23}{32}$ **c** 5 min 55 s

**2** **a** $k\int_3^7 x-3\,dx + k\int_7^{11} 11-x\,dx = 1$

$k\left(\left[\frac{x^2}{2}-3x\right]_3^7 + \left[11x-\frac{x^2}{2}\right]_7^{11}\right) = 1$

$16k = 1 \qquad k = \frac{1}{16}$ as required

**b**

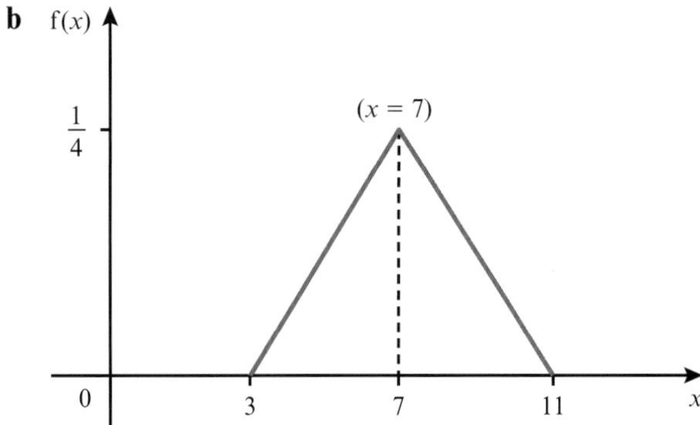


The sketched distribution is symmetrical about $x = 7$.

**c** $11-2\sqrt{2}$ or 8.17 **d** $\frac{8}{3}$

**3** **a** $\frac{1}{3}$

**b**

The graph is uniform.

**c** 45 min

**d** 0.286

**e** 0.0953

**4** **a** $\frac{14}{27}$

**b** Mean $= \frac{9}{8}$; var $= \frac{171}{320}$

**c** 0.6603 (Integrate from 0 to $\frac{10}{7}$.)

**5** **i** $X, Z, Y, W$ or $\sigma_X, \sigma_Z, \sigma_Y, \sigma_W$

**ii** **a** Mean = 0

$\frac{1}{18}\int_{-3}^{3} x^4\,dx - 0$

$= \frac{1}{18}\left[\frac{x^5}{5}\right]_{-3}^{3}$

$= 5.4$

$\sqrt{5.4} = 2.32$, as required.

**b** 0.268

**c** 0

**6** **a** 0.8

**b** **i** $\int_0^{1.5} k\left(2.25-x^2\right)dx = 1$

$k\left[2.25x-\frac{x^3}{3}\right]_0^{1.5} = 1$

$k = \frac{4}{9}$, as required.

**ii** 56 h 15 min

**iii** Max value of $x$ is 150 hours, whereas daffodils can last for 290 hours.

**iv** 2.9 to 5 inclusive

**7** **a** $\frac{5}{32}$

**b** **i** $\sqrt{\frac{2}{\pi}}$ or 0.798

**ii** 0

**iii** 0.8

**8** $\sqrt{8}$ or $2\sqrt{2}$ or 2.83

**9** **i** Longest lifetime

**ii** $\int_1^a \frac{k}{x^2}\,dx = 1$

$\left[\frac{-k}{x}\right]_1^a = 1$

$k = \frac{a}{a-1}$, as required.

**iii** 1.53

**10** **i** $\int_0^{\frac{2\pi}{3}} k\sin x\,dx = 1$

$\left[-k\cos x\right]_0^{\frac{2\pi}{3}} = 1$

$k = \frac{2}{3}$, as required.

**ii** $\frac{2}{3}\int_0^m \sin x \, dx = 0.5$

$\frac{2}{3}\left[-\cos x\right]_0^m = 0.5$

$\frac{2}{3}(-\cos m + 1) = 0.5$

$\cos m = 0.25$

$m = 1.32$

**iii** 1.28 to 3 s.f.

**11 i** $\int_{-1}^{1} k(1-x)\,dx = 1$

$k\left[x - \frac{x^2}{2}\right]_{-1}^{1} = 1$

$k = \frac{1}{2}$, as required.

**ii** 0.0625

**iii** $-\frac{1}{3}$

**iv** $1 - \sqrt{3}$ or $-0.732$

**12 a** $\int_3^6 k(6t - t^2)\,dt = 1$

$k\left[3t^2 - \frac{t^3}{3}\right]_3^6 = 1$

$k = \frac{1}{18}$, as required.

**b** $\frac{33}{8}$

**c** $\frac{4}{27}$

**d** Less than 5 min, as answer to part **c** $< \frac{1}{4}$.

# 5 Sampling

## Prerequisite knowledge

**1 a** 0.894 **b** 0.290

**2 a** 0.807 **b** 0.0745

## Exercise 5A

**1 a** 713 299 680 324 413 623

**b** 039 282 237 992 325 799

**2 a** Reads across each row then down the column.

**b** Reads two values in each line.

**c** Reads down the column.

**3 a** This method would give only those people who arrive early to work the chance of being selected. It would also exclude anyone working from home, or anyone who is away for some reason, such as on holiday or sick.

**b** Starting with second group of three, taking last value of that group with first value of next group of three.

**4 a** These people have already made the decision to buy and so are unlikely to be representative of people in general.

**b** This restricts the age range from which the sample can be selected. People outside the small age range of 25 to 29 years do not have a chance of being selected.

**c** Only men have a chance of being selected.

**5** The different totals have different probabilities; not all students have the same chance of being chosen.

**6** Number employees from 001 to 712. Generate 3-digit random numbers. Ignore 000 and numbers above 712. Ignore numbers which have come up previously. Repeat the process until 50 distinct employees have been selected.

## Exercise 5B

**1** $\bar{X} \sim N\left(6, \frac{8}{80}\right)$; probability = 0.897

**2** $\bar{X} \sim N\left(30, \frac{36}{100}\right)$; probability = 0.0478

**3** $\bar{X} \sim N\left(21, \frac{4.2^2}{50}\right)$; probability = 0.954

**4** 0.0548

**5** 0.0019

**6 a** 0.0031 **b** 0.974

**7 a** 0.990 **b** 0.508

**8** 0.846

## End-of-chapter review exercise 5

**1 a i** 0.0049 **ii** 0. 932

**b** Original distribution is normal since $n$ is only 20.

**2 a** $E(X) = 3.7$

$\text{Var}(X) = 2^2 \times 0.1 + 3^2 \times 0.4 + 4^2 \times 0.2 + 5^2 \times 0.3 - 3.7^2 = 1.01$, as required.

**b** $S \sim N(370, 101)$

**c** 0.0207; central limit theorem means $S$ is approximately normal.

**3** 0.924

**4 a** 0.991 **b** 15.3

**5 a** Not all sums of scores have the same probability.

**b** Any valid method; random number tables, pieces of paper, Excel function, etc.

**6** **i** 59

**ii** Any value, to three significant figures, from 0.687 to 0.693 inclusive.

**iii** Repeats may be included, so another random number would have to be chosen.

**7** **i** $\bar{N} \sim N(352, 2.9)$ **ii** 0.120

**8** **i** Likely to be friends if sitting at the same table; only considers people who use the canteen.

**ii** 2-digit numbers; ignore > 82; ignore repeats.

**9** **i** The sample includes only readers of that issue; it includes only readers who care sufficiently to complete and return the questionnaire.

**ii** 4975 3952 (0)386 or 4975 5203 6088

**10** **i** $N\left(48.8, \frac{15.6^2}{5}\right)$ **ii** 0.568

# 6 Estimation

## Prerequisite knowledge

1 1.5, 0.1, 0.316

2 54.3, 386, 19.6

3 20, 12.9, 3.59



4 13, 23.1, 4.81

5 $H_0: \mu = 86$, $H_1: \mu \neq 86$, $\Phi\left(\frac{84-86}{4}\right) < 5\%$

6 $H_0: \mu = 54$, $H_1: \mu < 54$, $\Phi\left(\frac{50-54}{3}\right) < 5\%$

7 $H_0: \mu = 18$, $H_1: \mu > 18$, $\Phi\left(\frac{20-18}{\sqrt{3}}\right) > 99\%$

8 $X \sim N(16.8, 10.08)$

9 $X \sim N(55, 24.8)$

## Exercise 6A

1 $\bar{x} = 2.2$; $s^2 = 0.667$

2 $\bar{t} = 21$; $s^2 = 18.6$

3 $\bar{x} = 4.2$; $s^2 = 0.143$

4 $\bar{x} = 752$; $s^2 = 23.1$

5 $\bar{x} = 8.03$; $s^2 = 7.15$

6 $\bar{x} = 1.1$; $s^2 = 1.55$

7 $\bar{x} = 2.41$; $s^2 = 1.27$

## Exercise 6B

**1** $H_0: \mu = 18.4$; $H_1: \mu > 18.4$

$$\Phi\left(\frac{19.7-18.4}{\frac{3.6}{\sqrt{40}}}\right)$$

$\Phi(2.284) = 0.9888$

$1.12\% > 1\%$

Accept $H_0$. There is no evidence to suggest the tablets do not work that quickly.

**2** $H_0: \mu = 100$; $H_1: \mu > 100$

**a** $\bar{x} = 102.4$

$s^2 = 219.4$

$$\Phi\left(\frac{102.4-100}{\sqrt{\frac{219.4}{180}}}\right)$$

$\Phi(2.174) = 0.9852$

$1.48\% < 2\%$

Reject $H_0$ and accept $H_1$. There is some evidence to suggest the mean IQ is higher than 100.

**b** $\bar{x} = 104$

$$\Phi\left(\frac{104-100}{\frac{15}{\sqrt{6}}}\right)$$

$\Phi(0.653) = 0.7432$

$25.68\% > 2\%$

Accept $H_0$. There is insufficient evidence to suggest the mean IQ is higher than 100.

**c** Results obtained from a small sample may not be as reliable as those obtained from large samples.

**3** $H_0: \mu = 380$; $H_1: \mu < 380$

$$\Phi\left(\frac{378.7-380}{\sqrt{\frac{6.4}{10}}}\right)$$

$\Phi(-1.625) = 0.0521$

$5.21\% > 5\%$

Accept $H_0$.

Assume variance unchanged. There is insufficient evidence to suggest the mean mass of pesto has reduced

**4** $H_0: \mu = 68$; $H_1: \mu < 68$

$\bar{x} = 66.625$

$$\Phi\left(\frac{66.625-68}{\frac{1.7}{\sqrt{8}}}\right)$$

$\Phi(-2.288) = 0.011$

$1.1\% > 1\%$

Accept $H_0$.

Assume egg masses are normally distributed and variance unchanged. There is no evidence to support the claim that the eggs are underweight.

5 $H_0$: $\mu = 80$; $H_1$: $\mu < 80$

$\bar{x} = 77.5$

$\Phi\left(\frac{77.5 - 80}{\sqrt{\frac{9}{6}}}\right)$

$\Phi(-2.041) = 0.0207$

$2.07\% < 5\%$

Reject $H_0$.

Assume variance unchanged. There is some evidence to suggest the amount of ice cream is too low.

6 $H_0$: $\mu = 2$; $H_1$: $\mu < 2$

$\bar{x} = 1.97675$

$s^2 = 0.01887$

$\Phi\left(\frac{1.97675 - 2}{\sqrt{\frac{0.01887}{80}}}\right)$

$\Phi(-1.514) = 0.065$

$6.5\% > 5\%$

Accept $H_0$. There is no evidence to suggest the bags of potatoes are underweight.

7 $H_0$: $\mu = 2000$; $H_1$: $\mu \neq 2000$

$\bar{x} = 1997.5$

$s^2 = 234.66$

$\Phi\left(\frac{1997.5 - 2000}{\sqrt{\frac{234.66}{42}}}\right)$

$\Phi(-1.058) = 0.145$

$14.5\% > 5\%$

Accept $H_0$.

Assume the lifetime of bulbs follows a normal distribution. There is no evidence to suggest the average lifetime of bulbs is different.

## Exercise 6C

1 a (1.86, 2.54) b (1.96, 2.44)

2 a (19.5, 22.5) b (19.2, 22.8)

3 a (3.88, 4.52, 18.2, 20.1) b (18.4, 19.9)

4 a (1.94, 2.00) b (1.93, 2.02)

5 a (7.55, 9.41) b 92.3%

6 a $\bar{x} = 1.95$; $s^2 = 0.0775$ b (1.87, 2.03)

c 90%

7 a $2 \times 2.236 = 4.472$ b 49

## Exercise 6D

1 a (0.140, 0.260) b (0.115, 0.285)

2 (0.308, 0.428)

3 (0.105, 0.345)

4 a (0.507, 0.581) b 9

5 45

6 a (0.127, 0.273)

b (0.263, 0.437). Since the confidence intervals intersect, from 0.263 to 0.273, at the 99% confidence level it is possible that more than 35% of bees in the colony are infected, and so it might collapse.

## End-of-chapter review exercise 6

1 a (0.0584, 0.142)

b i $0.113 + \frac{1}{2}(0.207 - 0.113)$ or $0.207 - \frac{1}{2}(0.207 - 0.113)$

ii 80%

2 a Average needs to be higher than contents as the sold product cannot be less than that stated.

b This is a small sample so we have to assume that the mass is normally distributed. $H_0$: $\mu = 276$; $H_1$: $\mu \neq 276$

$\Phi\left(\frac{277.7 - 276}{\frac{1.8}{\sqrt{8}}}\right)$

$\Phi(2.671) = 0.9962$

$99.62\% > 97.5\%$

Reject $H_0$.

Assume variance unchanged.

3 a (19 600, 22 400)

b i $1 - 2 \times 0.001 = 99.8\%$ ii \$23 194

4 a $\bar{x} = 288$; $s^2 = 650$ b (283, 293)

5 a $\bar{x} = 81.5$

$H_0$: $\mu = 80$; $H_1$: $\mu > 80$

$\Phi\left(\frac{81.5 - 80}{\frac{2.6}{\sqrt{8}}}\right)$

$\Phi(1.632) = 0.9486$

$94.86\% < 95\%$

Accept $H_0$. There is insufficient evidence to reject the claim.

Assume sd unchanged. Battery charging times are normally distributed.

**b** $H_0: \mu = 24$; $H_1: \mu < 24$

$\Phi\left(\dfrac{23.2 - 24}{\frac{1}{\sqrt{5}}}\right)$

$\Phi(-1.789) = -0.9633$

$3.67\% < 5\%$

Reject $H_0$.

Assume variance unchanged.

**c** 26.0 h

**6** 11, 12, 13, 14 or 15

**7** **i** $H_0: \mu = 2.60$

$H_1: \mu > 2.60$

$\Phi\left(\dfrac{2.64 - 2.6}{\frac{0.2}{\sqrt{75}}}\right) = \Phi(1.732)$

Reject $H_0$. There is sufficient evidence to suggest the mean has increased.

**ii** 0.0345

**8** **i** One-tailed as the test refers to a possible increase.

**ii** 1.481 to $1.503 < 1.645$

**iii** Type II possible as $H_0$ not rejected.

**9** **i** $\dfrac{\bar{x} - 22}{\frac{3.5}{\sqrt{12}}} > 1.645$

**ii** 0.0172

**10** **i** (7.54, 9.26)

**ii** No, since the distribution of the population is normal.

**iii** 8 g lies within the confidence interval, so the claim is justified.

**11** **i** (29.3, 29.9)

**ii** 30 g lies outside the confidence interval, so the claim is not supported.

**iii** The confidence interval is a variable.

**12** **i** 0.145; $n = 600$ **ii** 97.4

**13** **i** There is insufficient evidence to conclude that the mean is less than 750 g.

**ii** $n = 18$

## Cross-topic review exercise 2

**1** $\displaystyle\int_1^t \frac{x}{8}\,dx = \int_1^t \frac{9}{4x^3}\,dx$

$\left[\dfrac{x^2}{16}\right]_0^t = \left[\dfrac{-9}{8x^2}\right]_1^t$

$\dfrac{t^2}{16} = \dfrac{-9}{8t^2} + \dfrac{9}{8}$

$t^4 - 18t^2 + 18 = 0$

$t^2 = 9 \pm 3\sqrt{7}$

$t = 1.03$ is the only possible solution since other solutions outside range for $t$.

**2** **a**

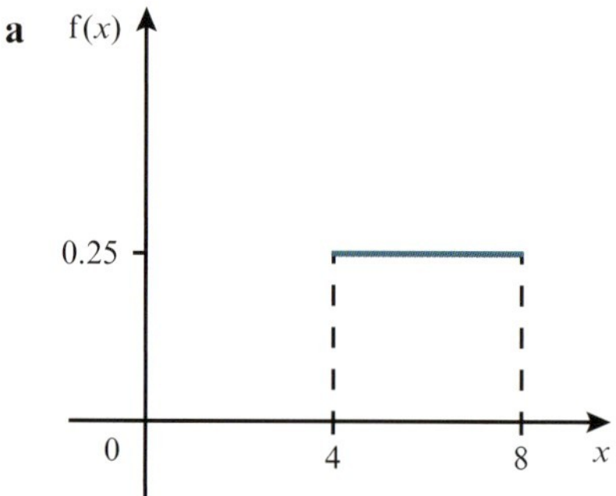


**b** Mean = 6; variance = $\dfrac{4}{3}$

**c** $\bar{X} \sim N\left(6, \dfrac{1}{30}\right)$

**d** $a = 6.23$

**3** **a** Mean = 14.167; variance = 7.639

**b** $\bar{T} \sim N(14.167, 0.382)$

**c** $14.167 - 2 \times \sqrt{0.382} = 12.93 > 12.4$

Alt.

$\Phi\left(\dfrac{12.4 - 14.167}{\sqrt{0.382}}\right) = \Phi(-2.86) = 0.0021$

**4** **i** Testing destroys items; it takes too long or is too expensive to try to test the population.

**ii** **a** (64.9, 66.5)

**b** 64.7 lies outside the confidence interval; therefore, the mean height of bounce probably is affected.

**5** **i** 2.09; 0.000131 **ii** 0.0832

**6** **i** Mean = 30.2; variance = 129

**ii** Mean = 55.2; variance = 291

**7** **i** (0.133, 0.307)

**ii** Mean = 127; variance = 11 001 or 11 000

**iii** Generate 4-digit numbers; ignore repeats; ignore values > 9526.

**8** **i** 0.146

**ii** **a** Looking specifically for a decrease.

**b** There is sufficient evidence to reject $H_0$; the mean has decreased.

**c** No, because the population is normally distributed.

**9** **i** Mean = 2866 or 2870; variance = 4130

**ii** There is no evidence to suggest that the mean distance has changed.

**10** **i** (56.6, 67.4)

**ii** 92% of all confidence intervals will contain the mean.

**iii** Each possible sample is equally likely to be chosen.

**11** **i** Assume that the standard deviation is unchanged. There is sufficient evidence to suggest that the mean weekly profit has increased.

**ii** Yes, because the distribution of $X$ is unknown.

**iii** 0.05

**iv** 0.0091

**12** **i** Sampling from a telephone directory excludes: people without a phone; those who are ex-directory; children; partners of named person.

**ii** (0.119, 0.261)

**iii** $x = 92.8\%$ (2 significant figures)

**13** **i** Mean = 331(.125); variance = 4.125 or 4.13

**ii** (330, 332)

**iii** No, because 333 does not lie within the confidence interval.

**14** **i** 0.941

**ii** $m < 2.53$ and $m > 2.71$

**15** **i** Multiplying by 2 means odd numbers are not included.

**ii** (1.54, 1.70)

**16** **i** 0.324

**ii** 0.994

**17** $H_0$: $\mu = 42$; $H_1$: $\mu \neq 42$. The teacher's estimate can be accepted.

**18** **i** $\mu = 4.27$; $\sigma^2 = 0.00793$

**ii** (4.25, 4.29)

**iii** 9

**19** **i** $\int_1^4 kt^{\frac{1}{2}}\,dt = 1$

$$\left[\frac{2kt^{\frac{3}{2}}}{3}\right]_1^4 = 1$$

$$k = \frac{3}{14}$$

**ii** 2.66 h **iii** 2.73 h **iv** 0.0243

**20** **i** Only train travellers are asked the question, so very many adults are excluded.

**ii** People travelling to work on this train.

**iii** Mean = 6.17; variance = 0.657

## Practice exam-style paper

**1** **a**

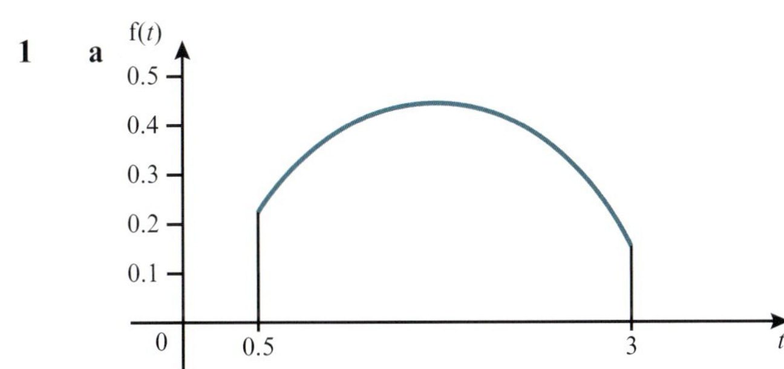


$$k\int_{0.5}^{3} 10t - 3t^2\,dt = 1$$

$$k\left[5t^2 - t^3\right]_{0.5}^{3} = 1$$

$$k = \frac{8}{135}$$

**b** **i** $\frac{8}{135}\left[5t^2 - t^3\right]_2^3 = \frac{16}{45}$

**ii** $B\left(10, \frac{16}{45}\right)$

$${}^{10}C_0 \frac{16}{45}^0\left(1 - \frac{16}{45}\right)^{10} + {}^{10}C_1 \frac{16}{45}^1\left(1 - \frac{16}{45}\right)^9$$

$$= 0.0805$$

**c** $\frac{8}{135}\int_{0.5}^{3} 10t^2 - 3t^3\,dt = \frac{8}{135}\left[\frac{10t^3}{3} - \frac{3t^4}{4}\right]_{0.5}^{3} = 1.711$

**d** Not realistic, as it is unlikely that no calls will be connected within $\frac{1}{2}$ a minute.

**2** **a** **i** $e^{-0.9}\frac{0.9^3}{3!} = 0.0494$

**ii** $0.0494^3 = 0.000121$

**iii** $e^{-0.9}\left(1 + 0.9 + \frac{0.9^2}{2}\right) = 0.937$

**b** $e^{-(4 \times 0.9)} = 0.0273$

**c** $H_0$: $\lambda = 3.6$; $H_1$: $\lambda < 3.6$;

$e^{-3.6}(1 + 3.6) = 0.126$

As this is greater than 10%, accept $H_0$. There is insufficient evidence to say that the number of breakdowns has decreased.

**3** $B\left(12, \frac{1}{2}\right)$

$$\binom{12}{0}\frac{1}{2}^{12} + \binom{12}{1}\frac{1}{2}^{12} + \binom{12}{2}\frac{1}{2}^{12} = 0.0193$$

Minimum level of significance = 2%

**4** $8 \times 17 + 8 \times 4.5 + 50 = 222$

$8 \times 2^2 + 8 \times 0.3^2 = 32.72$

$\Phi\left(\frac{230 - 222}{\sqrt{32.72}}\right)$

$= \Phi(1.399) = 0.919$

**5** **a** $32.2 \pm z \times 3.9$

$z = 1.96$

Required interval $= (24.6, 39.8)$

**b** $\frac{301.6}{10} = 30.16$

$30.16 \pm z \times \frac{3.9}{\sqrt{10}}$

$z = 1.645$

Required interval $= (28.1, 32.2)$

**6** Assume that the lifetime of the batteries is normally distributed.

$H_0$: $\mu = 76$; $H_1$: $\mu \neq 76$

$\Phi\left(\frac{74.6 - 76}{\frac{4}{\sqrt{40}}}\right)$

$\Phi(-2.214) = 0.0135$

$1.35\% > 1\%$

Accept $H_0$; there is insufficient evidence to reject the manufacturer's claim.

# Glossary

## A

**Acceptance region:** the area of the graph, or set of values, for which you accept the null hypothesis

**Alternative hypothesis, $H_1$:** The hypothesis that the observations being considered in a hypothesis test are influenced by some non-random cause; the abbreviation for the alternative hypothesis is $H_1$

## B

**Bias:** is the tendency of a statistic to overestimate or underestimate a parameter

**Binomial distribution:** a discrete distribution where there is a fixed number, $n$, of trials with only two outcomes, success or failure, where the probability of success is $p$

## C

**Central limit theorem:** states that, provided $n$ is large, the distribution of sample means of size $n$ is $\bar{X}(n) \sim N\left(\mu, \frac{\sigma^2}{n}\right)$, where the original population has mean $\mu$ and variance $\sigma^2$. The central limit theorem allows you to use the normal distribution to make statistical judgements from sample data from any distribution

**Confidence interval (CI):** an interval that specifies the limits within which it is likely that the population mean will lie

**Continuous random variable:** the outcome of a continuous random variable can take an infinite number of possible values; these are usually measurements

**Critical region (or rejection region):** the area of a graph, or set of values, for which you reject the null hypothesis. The boundary of the critical region is the **critical value**

**Critical value:** *see* **critical region**

## D

**Discrete random variable:** has outcomes that are countable distinct values

## H

**Hypothesis:** a claim believed or suspected to be true

**Hypothesis test:** the investigation to find out if a claim could happen by chance or if the probability of it occurring by chance is statistically significant

## N

**Normal distribution:** a continuous random distribution used to describe many naturally occurring phenomena. Its graph is a bell-shaped curve. It is used to describe distributions with unknown parameters. *See also* **central limit theorem**

**Null hypothesis, $H_0$:** the assumption that there is no difference between the parameter being tested and the sample data. *See also* **alternative hypothesis**

## O

**One-tailed hypothesis test:** in this test, the alternative hypothesis looks for an increase or decrease in the parameter. The critical region is at one end, or tail, of the graph of probabilities

## P

**Parameter:** a value that summarises data for a population

**Poisson distribution:** a discrete distribution where the number of occurrences of an event occur independently, at random and at a constant average rate

**Population:** refers to the complete collection of items in which you are interested

**Probability density function (PDF):** a function that describes the probabilities of a continuous random variable; its graph is never negative

**Probability distribution:** describes the probabilities associated with each possible value of a discrete random variable

## R

**Random sampling:** means that every possible sample of a given size is equally likely to be chosen. A random sample of size $n$ is selected in such a way that all possible samples of size $n$ that you could select from the population have the same chance of being chosen

**Random variable:** describes the numerical outcomes of a situation

**Rejection region:** *see* **critical region**

## S

**Sample:** a part of a population. A sample can be small or large

**Sampling frame:** a list of the whole population

**Sample mean, $\bar{X}$:** an estimate of the population mean, calculated from a group of observations

**Sampling error:** the error from choosing an unrepresentative sample

**Selection bias:** the unintentional selection of one group or outcome of a population over potential other groups or outcomes of the population

**Significance level:** the percentage level at which the null hypothesis is rejected. Typically the significance level used is 5% or 1%

**Standard error:** the standard deviation of sample mean, $\frac{\sigma}{\sqrt{n}}$

**Statistic (also known as an estimate):** a numerical value calculated from a set of data and used in place of an unknown parameter in a population

## T

**Test statistic:** the calculated value in a hypothesis test

**Two-tailed hypothesis test:** in this test, the alternative hypothesis looks for a difference from the parameter. The critical region is shared between both tails of the graph of probabilities

**Type I error:** a Type I error is said to have occurred when a true null hypothesis is rejected

**Type II error:** a Type II error is said to have occurred when a false null hypothesis is accepted

# Index

acceptance region 5
alternative hypothesis 6, 7

Bernoulli, Jacob 28
bias of an estimate 122
biased sampling 102
binomial distribution 26
    approximation to normal distribution 10–12
    Poisson approximation to 36–40
    population proportions 138
Bortkiewicz, Ladislaus 26

census data 100
    collation and analysis 101
central limit theorem 100, 110–15
class intervals 76
communication vi
confidence intervals 131–2
    effect of sample size 133–4
    for population mean
        normal populations with known variance 132–5
        using a large sample 135–6
    for population proportion 138–41
continuity corrections 10, 41
    and central limit theorem 115
continuous random variables 75
    calculating probabilities 79–82
    expectation and variance 88–90
    histograms 76–8
    median and other percentiles 83–6
    probability density functions 78–9
    rectangular (uniform) distribution 92
critical region 5, 8
critical values 5

data analysis 101
data collection 100
difference of independent random variables 56–9
discrete probability distributions 26
    *see also* Poisson distribution
DNA evidence 2, 3
drug trials 3

*e* 28
errors
    causes of 18
    Type I and Type II 17–20, 45

estimated parameters, notation 122
estimation 121–2
    bias of an estimate 122
    confidence intervals for population mean 131–2
        normal populations with known variance 132–5
        using a large sample 135–6
    confidence intervals for population proportion 138–41
    hypothesis testing of the population mean 127
        with known variance 127–9
        using a large sample 129–30
    reliability of 122
    unbiased estimates of population mean and variance 122–6
Euler, Leonard 28
expectation
    of continuous random variables 88–90
    of multiples of independent random variables 57–8
    of random variables and constants 51–5, 59–60
    of sample means 109
    of sum or difference of independent random variables 56–9
    *see also* mean

falsification theory, Karl Popper 5
Fisher, Ronald 3
frequency density 77

Gallup, George 139
geometric distribution 26

histograms 76–8
hypotheses 6
hypothesis testing
    approximating distributions 10–12
    conclusions 12
    critical region (rejection region) 5, 8
    critical value 5
    dice experiment 3–5
    left-handedness example 8
    null and alternative hypotheses 6, 7
    one-tailed and two-tailed tests 14–16
    pentagon spinner example 9–10
    with the Poisson distribution 43–5
    of the population mean 127
        with known variance 127–9
        using a large sample 129–30
    real-life applications 2–3
    significance level 5, 10
    simulator training example 6–7
    test statistics 8
    Type I and Type II errors 17–20, 45

interquartile range 84–5

linear combinations of random variables
    multiples of independent random variables 57–8
    with normal distributions 61–3
    with Poisson distributions 64–5
    random variables and constants 51–5, 59–60
    real-life applications 51
    sum and difference of independent random variables 56–9

mean
    confidence intervals for population mean 131–6
    of continuous random variables 88–90
    hypothesis testing of the population mean 127–30
    of multiples of independent random variables 57–8
    of a Poisson distribution 27–8
    of random variables and constants 51–5, 59–60
    of sample means 109
    of sum or difference of independent random variables 56–9
    unbiased estimates of population mean 122–6
    *see also* sample means
median 83–6
modelling vii
multiples of independent random variables 57–8

normal distributions 156
    approximation to the Poisson distribution 40–2
    binomial distribution approximation to 10–12

normal distributions (*Cont.*)
- central limit theorem 112–15
- critical values 157
- linear combinations of random variables 61–3
- standard normal table 156

null hypothesis 6, 7

one-tailed hypothesis tests 14–16
opinion polls 2, 3, 138–9

parameters 7
percentiles 83–6
Poisson, Simeon-Denis 36
Poisson distributions
- adaptation for different intervals 33–5
- approximation by the normal distribution 40–2
- approximation to the binomial distribution 36–40
- consumer hotline example 30–1
- hypothesis testing 43–5
- linear combinations of random variables 64–5
- mean and variance 27–8
- probability formula 28–9
- real-life applications 26–7
- typesetter errors example 31

Popper Karl, falsification theory 5
population mean
- confidence intervals 131–2
  - normal populations with known variance 132–5
  - using a large sample 135–6
- hypothesis testing 127
  - with known variance 127–9
  - using a large sample 129–30

population parameters, notation 122
population proportion, confidence intervals 138–41
population statistics 122
- unbiased estimates 122–6

populations 100
probabilities
- with continuous random variables 79–82
- dice rolls 3–4

probability density 77
probability density functions (PDFs) 78–9
- calculating probabilities 79–82
- expectation and variance 88–90
- median and other percentiles 83–6
- rectangular (uniform) distribution 92

problem solving vi

quality control 135
quartiles 84–6

random number generators and tables 103–5
random sampling 102, 103–5
- quality control 135

random variables
- multiples of 57–8
- sum and difference of 56–9
- *see also* continuous random variables; linear combinations of random variables

rectangular (uniform) distribution 92
rejection region 8

sample characteristics 102
sample means 122
- central limit theorem 110–15
- distribution of 106–10

sample sizes 100
- and central limit theorem 111–13
- effect on confidence intervals 133–4

sampling 100
- quality control 135
- real-life applications 100, 121
- use of random numbers 103–5

sampling error 122
sampling frame 103
sampling techniques 102–3
selection bias 102
significance levels 5
standard error 127
- and confidence intervals 132–3

statistics 122
sum of independent random variables 56–9

test statistics 6, 8
two-tailed hypothesis tests 14–16
Type I and Type II errors 17–20, 45

unbiased estimates 122–6
undergroundmathematics website vii
uniform (rectangular) distribution 92

variance
- of continuous random variables 89–90
- of multiples of independent random variables 57–8
- of a Poisson distribution 27–8
- of a population 122
- of random variables and constants 51–5, 59–60
- of sample means 109
- of sum or difference of independent random variables 56–9
- unbiased estimates of population variance 122–6